Communications
in Computer and Information Science　　2437

Series Editors

Gang Li ⓘ, *School of Information Technology, Deakin University, Burwood, VIC,*
Australia
Joaquim Filipe ⓘ, *Polytechnic Institute of Setúbal, Setúbal, Portugal*
Zhiwei Xu, *Chinese Academy of Sciences, Beijing, China*

AF166705

Rationale

The CCIS series is devoted to the publication of proceedings of computer science conferences. Its aim is to efficiently disseminate original research results in informatics in printed and electronic form. While the focus is on publication of peer-reviewed full papers presenting mature work, inclusion of reviewed short papers reporting on work in progress is welcome, too. Besides globally relevant meetings with internationally representative program committees guaranteeing a strict peer-reviewing and paper selection process, conferences run by societies or of high regional or national relevance are also considered for publication.

Topics

The topical scope of CCIS spans the entire spectrum of informatics ranging from foundational topics in the theory of computing to information and communications science and technology and a broad variety of interdisciplinary application fields.

Information for Volume Editors and Authors

Publication in CCIS is free of charge. No royalties are paid, however, we offer registered conference participants temporary free access to the online version of the conference proceedings on SpringerLink (http://link.springer.com) by means of an http referrer from the conference website and/or a number of complimentary printed copies, as specified in the official acceptance email of the event.

CCIS proceedings can be published in time for distribution at conferences or as post-proceedings, and delivered in the form of printed books and/or electronically as USBs and/or e-content licenses for accessing proceedings at SpringerLink. Furthermore, CCIS proceedings are included in the CCIS electronic book series hosted in the SpringerLink digital library at http://link.springer.com/bookseries/7899. Conferences publishing in CCIS are allowed to use Online Conference Service (OCS) for managing the whole proceedings lifecycle (from submission and reviewing to preparing for publication) free of charge.

Publication process

The language of publication is exclusively English. Authors publishing in CCIS have to sign the Springer CCIS copyright transfer form, however, they are free to use their material published in CCIS for substantially changed, more elaborate subsequent publications elsewhere. For the preparation of the camera-ready papers/files, authors have to strictly adhere to the Springer CCIS Authors' Instructions and are strongly encouraged to use the CCIS LaTeX style files or templates.

Abstracting/Indexing

CCIS is abstracted/indexed in DBLP, Google Scholar, EI-Compendex, Mathematical Reviews, SCImago, Scopus. CCIS volumes are also submitted for the inclusion in ISI Proceedings.

How to start

To start the evaluation of your proposal for inclusion in the CCIS series, please send an e-mail to ccis@springer.com.

Ke-Wei Huang · Qi Cao · Ruidan Su

Editors

Financial Technology

5th International Conference, ICFT 2024
Singapore, September 23–25, 2024
Proceedings

 Springer

Editors
Ke-Wei Huang ⓘD
National University of Singapore
Singapore, Singapore

Qi Cao ⓘD
The University of Glasgow
Glasgow, UK

Ruidan Su ⓘD
Shanghai Jiao Tong University
Shanghai, China

ISSN 1865-0929 ISSN 1865-0937 (electronic)
Communications in Computer and Information Science
ISBN 978-981-96-3810-9 ISBN 978-981-96-3811-6 (eBook)
https://doi.org/10.1007/978-981-96-3811-6

Preface

We are pleased to present the proceedings of the 5th International Conference on Financial Technology (ICFT 2024), successfully held in Singapore on September 23–25, 2024. The conference served as a platform to foster research, innovation, and collaboration in the fields of Financial Technology (FinTech), Artificial Intelligence, and Machine Learning.

The conference featured insightful presentations by experts and industry leaders from around the world. We are deeply grateful to our distinguished Keynote Speakers, Siau Keng Leng (City University of Hong Kong, China) and Erik Cambria (Nanyang Technological University, Singapore), and distinguished invited speakers, Ke-Wei Huang (National University of Singapore, Singapore), Chi Seng Pun (Nanyang Technological University, Singapore), Steven Li (RMIT University, Australia), Thierry H. Brutman (EDDA Stock Finance, France), and Jaber Jemai (Higher Colleges of Technology, UAE). In addition, we extend our sincere thanks to the more than 20 presenters whose contributions enriched the discussions and knowledge shared throughout the event.

ICFT 2024 received a total of 35 papers via the Openconf system, of which 17 papers were selected for inclusion in these proceedings. The authors represent institutions and organizations from various countries, including China, Singapore, the USA, the UK, South Africa, Indonesia, Italy, the UAE, etc. Each submitted paper underwent a rigorous peer-review process, utilizing a single-blind review. Every paper received 3 reviews, ensuring the highest academic and professional standards. We are immensely grateful to the scientific committee members for their dedicated efforts in making this possible.

Our heartfelt thanks also go to all participants, presenters, and organizers whose hard work and enthusiasm contributed to the success of ICFT 2024. Their commitment ensured that the conference was an enriching and impactful experience for all involved.

We hope that the proceedings of ICFT 2024 will serve as a valuable resource, capturing the latest advancements and ideas in the FinTech field. We believe these contributions will inspire continued progress and innovation in financial technology and related disciplines.

We eagerly anticipate future opportunities for collaboration and knowledge exchange within the global FinTech and AI communities.

Ke-Wei Huang
Qi Cao
Ruidan Su

Organization

Conference Chair

Xiaowei Ding   Nanjing University, China

Program Committee Chair

Ke-Wei Huang   National University of Singapore, Singapore

Program Committee Co-chairs

Qi Cao   University of Glasgow, UK
Ruidan Su   Shanghai Jiao Tong University, China

Technical Program Committee Chair

Ye Luo   University of Hong Kong, China

Local Chair

Peng Liu   Singapore Management University, Singapore

Publicity Chair

Xue Zhao   Zhejiang Gongshang University Hangzhou
                                   College of Commerce, China

Technical Program Committee

Haneen Al-Khawaja   University of Zurich, Switzerland
Abeer Alkhwaldi   Mutah University, Jordan
Murad Al-zaqeba   Universiti Sains Islam Malaysia, Malaysia

Hanudin Amin	Universiti Malaysia Sabah, Malaysia
Afshin Ashofteh	NOVA University of Lisbon, Portugal
Mat Razali Noor Afiza	National Defence University of Malaysia, Malaysia
Vadim Azhmyakov	Universidad Central, Colombia
Felix Chan	Macau University of Science and Technology, China
Tianxiang Cui	University of Nottingham Ningbo China, China
Gonçalo Dos Reis	University of Edinburgh, UK
Yaoyao Fan	Soochow University, China
Yeli Feng	Amplify Health Asia Limited, Singapore
Ştefan Cristian Gherghina	Bucharest University of Economic Studies, Romania
Serna Calvo Gregorio Manuel	University of Alcalá, Spain
Ahmed Mohamed Habib	Accounting and Finance Independent Research, Egypt
Guanming He	Durham University, UK
Jaber Jemai	Higher Colleges of Technology, UAE
Rashid Khalil	Bahrain Polytechnic University, Bahrain
Mehdi Khashei	Isfahan University of Technology, Iran
Carol Anne Hargreaves	National University of Singapore, Singapore
Hsing-Hua Hsiung	Chaoyang University of Technology, Taiwan
Thi Le	Murdoch University, Australia
Gyu Myoung Lee	Liverpool John Moores University, UK
Wonjun Lee	Cheongju University, South Korea
Yi Man Li (Rita)	Hong Kong Shue Yan University, China
Chee Yoong Liew	UCSI University, Malaysia
Aijun Liu	Xidian University, China
Jose Liu	Newcastle University, UK
Ooi Kok Loang	City University Malaysia, Malaysia
José Luís Martins	Polytechnic Institute of Leiria, Portugal
Daniel Rabetti	National University of Singapore, Singapore
Siva Shankar Ramasamy	Chiang Mai University, Thailand
Ramona Rupeika-Apoga	University of Latvia, Latvia
Rafael Felipe Schiozer	Fundação Getulio Vargas's São Paulo School of Business Administration, Brazil
Leilei Shi	University of Science and Technology of China, China
Alexey Mikhaylov	Financial University under the Government of the Russian Federation, Russia
Adel Mohammed Yaslam Sarea	Ahlia University, Bahrain
Hongfei Tang	Seton Hall University, USA
Vu Trinh	Newcastle University, UK

Khaw Khai Wah	Universiti Sains Malaysia, Malaysia
Yaopeng Wang	University of Shanghai for Science and Technology, China
Yishun Wang	National Changhua University of Education, Taiwan
Kang Xin	Tokushima University, Japan
Xue Zhao	Zhejiang Gongshang University Hangzhou College of Commerce, China
Huanle Zhang	Shandong University, China
Chengli Zheng	Huazhong Normal University, China
Diana Zuhroh	Universitas Merdeka Malang, Indonesia

Contents

Fintech: Artificial Intelligence and Machine Learning Leverage 1
  Jaber Jemai

Improved Machine Learning Algorithms for Fraud Detection in Fintech
Companies 9
  Chukwuemeka Nwachukwu, Chukwuebuka Akwiwu-Uzoma,
  Samuel Ovuehor, and Kehinde Durodola-Tunde

A Cognitive Analysis of CEO Speeches and Their Effects on Stock Markets ... 20
  Rohan Manro, Rui Mao, Liza Dahiya, Yu Ma, and Erik Cambria

Does E-Money Mediate the Effect of Fundamental Factors on the Stock
Price Index? 32
  Diana Zuhroh, Rini Setyowati, Sihwahjoeni, Gaguk Apriyanto,
  and Abdul Malik

Multi-Attribute Decision Making (MADM) Model Based on Bayesian
Neural Network Classification with Comparative Scoring Quantification
Method Can Solve Intangible Value Valuation Issues in Benefit-Cost
Analysis (BCA) 45
  Zhaojie Wang and Hoi-Hei Wang

Exploring Vulnerabilities in Near Field Communication (NFC) Devices:
A Comprehensive Investigation 56
  Sphamandla Sangweni and Khutso Lebea

A Novel Electronic Payment System Based on Zero-Knowledge Proof
and Blockchain 67
  Viet-Thang Nghiem, Thi-Huong Tran, and Ba-Lam Do

A Comprehensive AI and Blockchain Framework for Detecting
and Preventing Money Laundering in Bangladesh Financial Systems 76
  Mohammed Mizanur Rahman and Maisha Karim

Detecting Persuasion in Financial Short Texts: A Computational Approach 88
  Rajdeep Kumar, Rudra Chandra Ghosh, Ganesh Bahadur Singh,
  and Nitin Sharma

The Impact of Colored Noise on the CIR Model 99
  A. Pavlova, G. Zotov, and P. Lukianchenko

A Secure NFC Cardless Cash Withdrawal System 111
  Min-Shiang Hwang, Cheng-Ying Lin, Chun-Hsien Chang,
  and Cheng-Ying Yang

FinTech Digital Transformation: Generative AI, Humanoid Robots,
Metaverse, Human-AI Collaboration, and Industry 5.0 122
  Yuxin Liu, Runyu Wang, and Keng Siau

Artificial Intelligence (AI) and Virtual Reality Convergence in Financial
Services: The Power of Digital Twin Robo-Advisers 137
  Marco I. Bonelli and Jiahao Liu

CITRONN: A Convolutional Neural Network for Crypto Image-Based
Trading 151
  Haiyun Zhu, Beier Liu, and Mingjun Sun

Can the Government's Distribution of Consumption Vouchers Stimulate
Adoption of Digital Payment Channels? Insights from a Social Learning
Perspective 164
  Roman Podkorytov, Ying Jen Chiang, and Ron Chi-Wai Kwok

Do Investor Protection Measures Matter for Equity Crowdfunding Portals? 176
  Li Jun Chen and Steven Li

Financial Time Series Simulation with Transformer-Based Generative
Models Under Continuous Conditions 191
  Horstann Rui Yao Ho and Chi Seng Pun

Author Index 207

Fintech: Artificial Intelligence and Machine Learning Leverage

Jaber Jemai$^{(\boxtimes)}$ (iD)

Higher Colleges of Technology, Abu Dhabi, UAE
`jjemai@hct.ac.ae`

Abstract. Financial Technology (Fintech) is currently an area of deep and continuous disruptions enabled by the rise of technological advancements notably from the telecommunications and information technology domains. Financial institutions are taking advantage of improving their offerings. Thus, as customers, we can see daily new services, enhanced platforms, customized procedures, among others. Artificial Intelligence (AI) and Machine Learning (ML) are the main enablers of Fintech reshape by revamping banking and insurance services. In this paper, we present the current and new abilities offered by AI and ML to financial institutions. We will mainly highlight the business applications that have been areas of deep improvement using AI and ML models and techniques like credit risk analysis, fraud detection, and smart algorithmic trading. While developing innovative AI/ML based models to address financial problems, it is important to overcome a multitude of challenges including data privacy, data accessibility, security, and the shortage of skilled AI/ML professionals.

Keywords: Fintech · Artificial Intelligence · Machine Learning

1 Introduction

Fintech is the application of modern technologies to financial products and services. It refers to a variety of technical improvements in financial services, including mobile banking, online lending platforms, digital payment systems, robot-advisors, and blockchain-based applications like cryptocurrencies [1]. Fintech companies include both new and established technological and financial firms that seek to enhance, supplement, or replace traditional financial services. The projected revenue of Fintech applications started scanting early last decade with $19 billion in 2015. However, it progressed exponentially to reach $33.3 billion in 2020 and to $245 billion in 2022. It is expected to reach $1.5 trillion in 2030 as reported in BCG group in 2023[1]. Moreover, Fintech holds currently 2% of the revenue of the global financial service market (Fig. 1), estimated at $12.5 trillion. It is projected to attain 25% in 2030 with an expected growth rate of 7%[2].

Geographically, the Fintech landscape covers all continents with high concentration in North America, the land of most technological advances. It is important to mention that

---

[1] BCG report, 'Reimagining the Future of Finance', May 2023.

[2] McKinsey report, 'Fintechs: A new paradigm of growth', October 2023.

K.-W. Huang et al. (Eds.): ICFT 2024, CCIS 2437, pp. 1–8, 2025.
https://doi.org/10.1007/978-981-96-3811-6_1

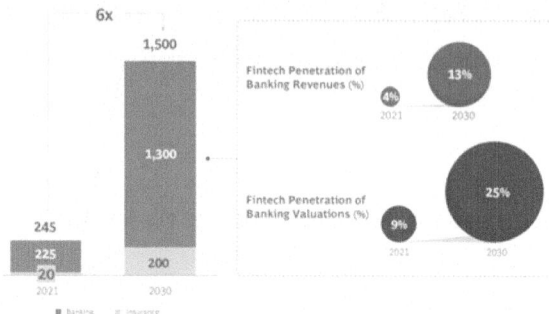

Fig. 1. Global Fintech revenue split by banking and insurance (BCG Report 2023).

the current big Fintech players are in the USA like Visa and PayPal. However, Fintech startups' dispersion is balanced with young companies emerging around the globe. We can see less startups in Africa and Latin America compared to other countries. Such evolution is catalyzed by technological advances arising in different disciplines like data and information sciences, telecommunications technology, quantum computing, artificial intelligence and machine learning, etc. In this paper, we will focus on the leverage offered by AI/ML in lifting-up Fintech services in terms of quality, spread, and value.

Artificial Intelligence contribution to Fintech covers various financial core business applications like risk assessment and management in credit, loans and mortgages, fraud detection and prevention, financial advising, smart trading, security and cyberattacks prevention, and services automation, customization, and availability. In this paper, we address various financial applications where AI and ML are offering a considerable competitive advantage to adoptive financial institutions against their competitors.

The paper is organized as follows; in the next section we present a contextual diagram summarizing fintech development ecosystem. Next section will detail some AI/ML technologies that can be widely applied to solve a big variety of problems on different disciplines. Section 4 will be devoted to the main AI/ML applications in Finance and Insurance with the highlight of the advantages gained by adoptive companies. In Sect. 5, we will study some challenges decelerating the pace of AI/ML applications in the financial arena. In the last section, we present some conclusions and research perspectives.

2   AI/ML in Fintech: A Contextual Diagram

Fintech is built up by the set of applications used by financial institutions to automate their backend and frontend services. By backend services, we refer mainly to applications used to collect, store, secure, and manage institutions' assets (hardware, software, data). Frontend services are composed of applications dealing with customer services like online banking and customer support. In the Figure below (Fig. 2), we illustrate the context of research and development of AI/ML applications to solve financial problems.

It is obvious that technological innovation has come to answer business needs that emerged at the lowest level of the figure. At the business layer, financial institutions raise issues related to the requirements to enhance some services and propose new products. As a result, the first answer comes always from the theoretical layer by consulting the available scientific body of knowledge in related areas. As for the scope of our study, computer science, information science, mathematics, and physics can offer some thesis and theories able to build the basis for innovative solution targeting raised business problems. That's by converting the established theories into technologies. To cite few, blockchain, Internet of Things (IoT), Radio Frequency Identification (RFID), Near-Field Communication (NFC) protocols, Biometric technologies, bigdata technology, and quantum computing. This asset of technologies represents an armada of candidate tools to drive the innovation phase and solve business concerns.

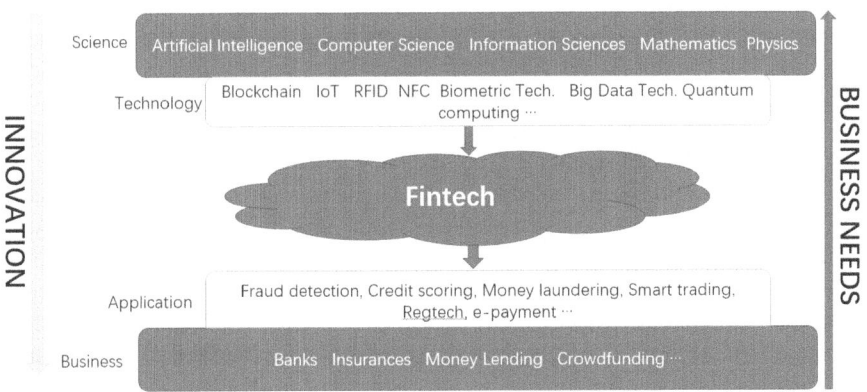

Fig. 2. AI/ML Innovation in the service of the financial business needs.

It is important to clarify that up to this point, the scope of application of the above-mentioned technologies is wide and can be instantiated to be used in other fields like healthcare, energy, social sciences, etc. In finance, many problems can be solved using such technologies like credit risk assessment and credit scoring for bankers, insurers, money lenders, etc. Similarly, fraud detection in credit cards or financial transactions can be solved using machine learning techniques. Anti-money laundering is currently one of the main research areas at theoretical and empirical levels. The ultimate objective is to propose, design, and develop AI/ML based solutions able to identify and track money laundering actors and transactions. Similarly, the trading field (stocks, forex, cryptocurrency) represents currently lucrative areas where AI/ML techniques can provide profitable trading strategies based on learning from technical and fundamental analysis as well as the design of generic trading strategies. In the remaining part of this paper, we will illustrate the above the leverage offered by AI/ML in solving the above-mentioned financial problems.

3 AI Technologies

Artificial intelligence offers a wide range of technologies able to provide smarter solutions for a big variety of fields including healthcare, robotics, finance, autonomous vehicles, among others. In finance, the below AI models have been extensively applied:

- Symbolic AI: this is the old school of AI. The objective is to build Expert Systems (ES) able to solve some basic problems. ES are composed mainly, by knowledge based built of explicit rules and a reasoning engine to resolve new questions using the knowledge base rules.
- Machine learning: by definition, ML consists of building models from data. Such models can discover insights from the training data and used later in classifying some objects, clustering observations into distinctive groups or autonomously drive a car using the punishment-reward principle. Considering the structure of the training set and the type of the training algorithm, three main types of ML models can be found: supervised learning where the dataset encloses the label of every observation. Unsupervised learning models identify implicit relationships between the observations of the training set, and able to group them into similar groups called clusters. However, the reinforcement learning consists of building AI smart agents using the punishment-reward principle.
- Generative AI: the Gen-AI models can build contents in response to user requests like text, images, videos, etc. established models like ChatGPT, Copilot, Synthesia, Canva, and Laland. Gen-AI models are basically artificial neural networks fed with big quantities of data during the training phase to identify some patterns and structures. Thereafter, during the test or usage phase they can create new and original artificial content. The field of Gen-AI is observing currently a real interest given the number of developed models for creating various contents and its high adoption by many industries. In fintech, Gen-AI can help improve financial operations by generating accounting and financial reports, handle financial planning and performance, and manage customers and investors relations.
- Natural Language Processing: NLP consists of equipping the machine (computer, mobile, tablet) with communication capabilities allowing them to exchange with Humans in natural language. NLP models can understand, interpret, summarize, and generate text in natural language. They are being used in various fields like sentiment analysis, opinion mining, chatbots, natural language translation, among others. In fintech, NLP powered chatbots and virtual assistants can provide financial advice to customers. Similarly, sentiment analysis can gauge public sentiment towards financial offerings.

4 Fintech Applications of AI/ML

AI and ML applications in financial markets are numerous. They help to enhance the operations efficiency, financial systems security, and improved customer experience. At the backend level, AI helps in improving procedures' automation, regulation compliance, decision support via deep data analytics and machine learning models. When dealing with customers, AI assists in designing more customized products. In this section, we will detail the main recent contributions of AI/ML in solving financial problems.

4.1 Fraud Detection

The detection of Credit Card Frauds (CCF) is a major challenge for banking institutions. Financial institutions must authorize each authentication transaction, making it possible for hackers to impersonate the cardholder and conduct fraudulent transactions. Companies like Visa and MasterCard had over 2.183 million cardholders [1]. The volume of loss due to fraudulent transactions on credit cards is more than 28.6 billion in 2020 and it is expected to reach $408 billion by the end of this decade.

Artificial intelligence and machine learning models have been extensively deployed to help in improving the detection of frauds. The CCF problem was modeled as a classification problem where the objective is to use the labelled training dataset to fit a classification model like a logistic regression [2], naïve bayes network [3], neural models [4], or ensemble models [5], among others. As a clustering problem, the CCF was investigated using K-nearest neighbors' approach [7] and the DBSCAN model [6]. The developed AI/ML models helped to find the attributes of fraudulent transactions to prevent as well as detecting the vulnerabilities in the authentication processes used by financial institutions [5].

4.2 Intelligent Trading Agents

The stock, forex, and cryptocurrencies markets are one of the main fields where financial institutions and individuals have a real stake. This is because the stakeholders can always gain more assets by employing profitable strategies in these markets. It is worth noting that the daily trading volume in forex is estimated to be $6.6 trillion. In this domain, AI and ML made a significant difference by the implementation of trading algorithms able to find identify more profitable positions after analyzing the current stock market or forex pair status. Stock prices forecasting [8] is one of the main areas of investigation using machine learning models like neural networks [9], deep learning [10, 11], XGBOOST [8], among others. In [12, 13], the authors investigated the design of profitable trading strategies. The authors reported an improved performance of the proposed ML based forecasting methods and trading strategies.

4.3 Anti-money Laundry

Money laundry consists of injecting money obtained from illegal sources in the legal financial networks to appear as it is coming from regular origin. Illegal sources of dirty money include all criminal activities like drug trafficking, human organs sale, human exploitation, terrorist funding, etc. The money collected is then credited to some illicit bank accounts or used to buy some real estate properties. The objective of anti-money laundry is to counter and track the illegal money actors and identify their used circuits. In [14], Gandhi et al. examined the potential offered by AI and ML to help in countering money laundry like deep learning [15] and neural networks [16].

4.4 Credit Risk Analysis

The main activities of financial institutions include money lending. Banks and lenders offer loans, credits, and mortgages to their customers. Lenders are expecting customers to pay back the full debt. However, it is not guaranteed that all customers will pay for the installments without defaulting. It is important for companies to minimize the loss given the default and its exposure to defaults. That's by carefully analyzing the credit risk before deciding on loan approval. Machine learning models are playing a major role in assisting lenders classifying their customers into potential defaulters or not [17]. Several ML models have been proposed including ensemble methods [18], deep learning [19], and neural networks [20].

5   Challenges of AI/ML Adoption in Fintech

The adoption of AI and ML models in financial applications bring guaranteed advantages for financial institutions. However, several challenges need to be overcome to gain a full benefice from AI models in fintech. Many of these obstacles are related to the nature of ML models as they are fully dependent on data. Below, we detail the principal barriers to AI adoption in fintech:

- Data Accessibility:it refers to the ease with which machine learning engineers and data scientists can get the datasets used to train ML models. ML models are fully dependent on data availability. In some contexts, particularly in financial institutions, real data cannot be made ready and in good quality for analysis and processing. In such a case, ML professionals make use of synthetic datasets to emulate real datasets [5]. Such alternatives seem to work fine, however their value and the insights that can provide cannot beat the value of those obtained from real datasets.
- Data quality: The datasets made available for analytics and processing using ML models need to meet certain criteria. That's to ensure valid training of ML models and to avoid any bias related to data quality issues. Datasets shall be accurate, complete, consistent, and valid. In financial context, relevant data can be collected from different sources including transactional databases, data warehouses, web, etc. Such a fact increases data veracity and lower data quality which will induce an invalid training of AI and ML models leading to erroneous findings.
- Privacy and security: The data manipulated by financial institutions is private in a big part. It is about information related to customers' accounts and banks assets. To be used for training machine learning models, a long legal and operational process should be undertaken to disclose the datasets. That's by getting the consent of the data owners, inform them how this information will be used, and set the usage limit. On another side, data owners need to be assured about the security of their data. Therefore, security enforcing mechanisms shall be in place like data encryption, access control, monitoring and auditing, authentication to guarantee proper usage of private data.
- Lack of skilled AI/ML manpower: Artificial intelligence and machine learning are relatively new and emerging disciplines. It was just recently that the universities started crafting and delivering fully fledged programs in artificial intelligence, machine learning and data science. This led to a gap between and increasing market demand for AI/ML professionals who can be part of the fintech industrial tissue. On the other hand, the applications of artificial intelligence cover too many fields. Such a situation led to the dispersion of the few skilled AI engineers over many domains.

6 Conclusions and Research Perspectives

Financial technology has significantly improved the quality of service rendered by financial institutions to their customers. It brought a high added value to banks, insurances, and money lending companies. However, the fintech disruption triggered by AI and ML led to an exponential leverage in the quality and value, and adoption of financial services by customers. Throughout various applications like credit risk analysis, fraud detection, smart trading bots, fighting money laundry, and tack trafficking networks, artificial intelligence-based systems showed a clear enhancement of financial operations. However, some factors might slow down the progress and adoption of AI perspective. Mainly, data related issues like accessibility, privacy, and security will harden the development of ML models. In this paper, we investigated the leverage offered by AI and ML to financial technology through the study of the main contribution and limits of AI-based models in solving difficult finance problems. This paper shed light on future applications of AI and ML in fintech. To cite a few, explainable AI can help in explaining the found insights and convince financial decision makers. We believe that the constructivism school of artificial intelligence has a lot to offer particularly the integration of expert systems and knowledge management.

References

1. Khang, A., et al. (eds.): Synergy of AI and Fintech in the Digital Gig Economy. CRC Press, FL (2024)
2. Saleh Alfaiz, N., Mohamed Fati, S.: Enhanced credit card fraud detection model using machine learning. Electronics 11(4), 662 (2022)
3. Varmedja, D., Karanovic, M., Sladojevic, S., et al.: Credit card fraud detection-machine learning methods. In: 2019 18th International Symposium INFOTEH-JAHORINA (INFOTEH), pp. 1–5. IEEE (2019)
4. Alarfaj, F.K., Malik, I., Khan, H.U., et al.: Credit card fraud detection using state-of-the-art machine learning and deep learning algorithms. IEEE Access 10, 39700–39715 (2022)
5. Jemai, J., Zarrad, A., Daud, A.: Identifying Fraudulent Credit Card Transactions Using Ensemble Learning. IEEE Access 12, 54893–54900 (2024)
6. Han, L.: Advanced dbscan: a clustering algorithm for personal credit reference system. Intelligent Systems and Applications:2019 IntelliSys 1, 370–381 (2020)
7. Awoyemi, J.O., Adetunmbi, A.O., Oluwadare, S.A.: Credit card fraud detection using machine learning techniques: A comparative analysis. In: 2017 international conference on computing networking and informatics (ICCNI), pp. 1–9. IEEE (2017)
8. El Mahjouby, M., et al.: Machine Learning Techniques for Predicting and Classifying Exchange Rates between US Dollars and Japanese Yen. Engineering, Technology & Applied Science Research 14(5), 16266–16271 (2024)
9. Navaei, M., Pahlevanzadeh, M.: Forecasting Next-Time-Step Forex Market Stock Prices Using Neural Networks. Advances in Machine Learning & Artificial Intelligence 5(2), 1–10 (2024)
10. Sevastjanov, P., Kaczmarek, K., Rutkowski, L.: A multi-model approach to the development of algorithmic trading systems for the Forex market. Expert Syst. Appl. 236, 121310 (2024)
11. Zhao, X., Huang, Y.: Analysing trends in trading patterns in financial markets using deep learning algorithms. Journal of Electrical Systems 20(3s), 1542–1555 (2024)

12. Al Ali, A:. Machine Learning Models for Enhanced Stock Trading Strategies. MS thesis. Rochester Institute of Technology (2024)
13. Chen, Y.: A study on automated improvement of securities trading strategies using machine learning optimization algorithms. Applied Mathematics and Nonlinear Sciences **9**(1)
14. Gandhi, H., et al.: Navigating the Complexity of Money Laundering: Anti–money Laundering Advancements with AI/ML Insights. International Journal on Smart Sensing and Intelligent Systems **17**(1) (2024)
15. Pazos, J.F.M., et al.: "Fraud Transaction Detection For Anti-Money Laundering Systems Based On Deep Learning. Journal of Emerging Computer Technologies **3**(1), 29–34 (2024)
16. Lokanan, M.E.: Predicting money laundering sanctions using machine learning algorithms and artificial neural networks. Appl. Econ. Lett. **31**(12), 1112–1118 (2024)
17. Jemai, J., Zarrad, A.: Feature Selection Engineering for Credit Risk Assessment in Retail Banking. Information **14**(3), 200 (2023)
18. Jemai, J., Chaieb, M., Zarrad, A.: A Big Data Mining Approach for Credit Risk Analysis. In: 2022 International Symposium on Networks, Computers and Communications (ISNCC). IEEE (2022)
19. Wang, J., et al.: Research on finance Credit Risk Quantification Model Based on Machine Learning Algorithm. Academic Journal of Science and Technology **10**(1), 290–298 (2024)
20. Emmanuel, I., Sun, Y., Wang, Z.: A machine learning-based credit risk prediction engine system using a stacked classifier and a filter-based feature selection method. Journal of Big Data **11**(1), 23 (2024)

Improved Machine Learning Algorithms for Fraud Detection in Fintech Companies

Chukwuemeka Nwachukwu[1], Chukwuebuka Akwiwu-Uzoma[2], Samuel Ovuehor[1], and Kehinde Durodola-Tunde[1(✉)]

[1] Department of Applied AI and Data Analytics, University of Bradford, Bradford, UK
k.durodola-tunde@bradford.ac.uk
[2] Department of Computing, University of Dundee, Dundee, UK

Abstract. The rapid adoption of mobile money is due to the penetration of smart devices and the reduced cost of computing power. Today, many rural dwellers use mobile smart devices (Ofcom, 2022). A financial solution that creates a transaction system through mobile smart devices is instrumental in closing the financial inclusion gap. Mobile money offers digital financial service providers access to an untapped market that is profitable. There has been a shift from traditional banking to digital banking services (mobile money). This is due to a rapid increase in mobile phone penetration and reliance on domestic remittances in many households (Mauree and Kohli, 2013). With the broad internet usage, modern technologies, and the current growth of digital financial service providers, there is a need for financial institutions to mitigate against financial crimes and fraudulent activities (Wewege et al., 2020). The sporadic expansion of digital transformation has made digital financial services providers vulnerable to cyber risks. The continuous partnerships with third parties, technology providers and fintech companies have heightened their vulnerability to cyber-attacks (Tsys, 2017). More people are prioritizing their delicate personal data against hacking, identity theft, cyberattacks and money laundering. Digital banks with lesser security infrastructure and small fintech companies are more prone to financial fraud and cyber risks. Hence, digital financial service providers and fintech companies should continuously update their security infrastructure (Wewege et al., 2020). This study focuses on how fintech companies can maximize improved machine learning algorithms for fraud detection.

Keywords: Machine Learning · Mobile Money · Fraud Detection

1 Introduction

Financial institutions have been consistently invaded by unscrupulous elements who are aimed at manipulating the financial ecosystem through criminal activities such as terrorism financing, identity fraud, financial fraud and money laundering. In 2022, approximately 4 million PayPal accounts were closed because fraudsters capitalized on their marketing programs to encourage mobile money among their customers. Fraudsters are reputed for enhancing their tricks and strategies to manoeuvre the security structure of a financial system. Therefore, financial organizations ought to know the fundamental behaviours and intentions of fraudsters. Mobile money is a terminology that covers various activities in the financial sector. It can be categorized into mobile commerce, mobile payment and mobile banking services (Atanu et al., 2014).

The uniqueness of a mobile money transaction is that it can be done on any smart device. This study focuses on mobile payment transactions in the digital remittance industry. Payment channels such as credit cards share similar risks as those in the digital remittance ecosystem. Nevertheless, there is a heightened risk in using mobile money transfer technologies because transactions happen swiftly. There are non-bank actors and more privacy than mobile commerce and mobile commerce platforms. Various studies on fraud detection have been carried out lately on the debit and credit card space. The heterogeneity and quantity of credit card transactions are highly important. Based on their currencies and locations, customers use credit cards for various activities. This shows a broader fraudulent scheme (Li et al., 2021). Given this, the essence of this research is to widen the scope of fraud detection study while integrating some insights from previous research to generate a fraud detection framework that would function excellently in the digital remittance space. The study aims to produce a fraud detection framework by making use of numerous supervised learning procedures to differentiate various algorithms and attain significant detection precision and accuracy. The purpose of this study is to evaluate the dataset and create an idea for a pre-processed application to systemize the data and generate the important information; to generate various machine learning categorization algorithms and examine their performance; to maximize cross-validation to recognize fraud detection algorithms suitable for fraud detection in any digital remittance firm. In the long term, the objective of the study is to produce a model and initiatives that use machine learning theories and analytics. The value of this study is reliant on the accuracy and precision of the categorization framework and how substantial the analysis is.

2   Money Transfer and Digital Remittance

The launch of mobile money transfer has recorded significant success in emerging economies with a huge population of middle and low-income earners (Asamoah et al., 2020). Digital remittance has been instrumental to the progress of digital financial inclusion and can be maximized by end-users, mobile network operators, service providers, partner banks, regulators and distribution channels. The goal is to create more inventive means of widening financial inclusion (Guermond, 2022). Digital remittances are made via payment tools in a self-assisted ecosystem or electronic medium. It is accepted by operational accounts such as mobile money accounts, microfinance accounts and bank accounts (Guermond, 2022). Digital remittance is cheaper, faster and a more suitable means of money transfer (IFAD, 2020). The shift from fiat currency to digital transaction is predicted to enhance the growth and expansion goals for financial inclusion by encouraging beneficiaries and migrants to save more boost their investments and maximize digital credit by linking digital remittance to a broader digital financial clime (GSMA, 2018). According to GSMA 2021, mobile money processed approximately $13 billion in global remittances. This is a 310 percent increase from the 2015 report and a 61 percent increase from the 2019 report.

2.1 Growth and Penetration of Mobile Money

Smart devices have widely redefined the activities and scope of the financial service industry. Although the extensive and penetrative usage of smart devices has been advantageous, it has created a chance for unscrupulous elements to manipulate. According to the GSMA 2022 Report, 4.2 billion people across the world use mobile internet. Mobile money has evolved in the previous decade from a subset financial service to a modern financial service that aided the economies of LMICs' shift from an unstructured financial system to a diverse and open digital economy (Awanis et al., 2022). In the last decade, the mobile money ecosystem has witnessed considerable development such as merchant payment, bill payments, international remittances and bulk disbursement. This is a visible signal of acceptance of mobile technology in the financial sector and the readiness of providers to vary.

During the COVID-19 crisis, more diasporas globally accepted mobile money channels as a better alternative to sending money to their friends and families. A 46 percent rise (approximately $16 billion) in mobile money international remittance was documented. This accounts for 3 percent of global international remittances (Awanis et al. 2022). Mobile money is not broadly accepted in some parts of the globe due to some regulatory interventions and government measures like costly data localization directives, irregular swift payment systems and transaction fees. Expensive costs of regulatory compliance can also hinder businesses from their main goal of offering low-cost financial solutions to their customers (Awanis et al., 2022).

2.2 Mobile Money Transfer and Digital Remittance Business Model

To create value, there is a need for a business model for a firm (Lurie, 2010). Every stakeholder is important in the implementation of the right business model. The business model can be bank-centered, MNO-centered or All-Inclusive (concerted) (Atanu et al. 2014). As time goes on, the business model becomes dynamic. Business models should be evolving depending on the existing business level, current regulatory procedures and operating ecosystem (Stein, 2011). Due to the complex fundamental system of connecting the telecom and banking sectors to form mobile money, there is a need to establish new business models. However, there is a need for continuous partnerships between every key player in the sector.

2.3 Mobile Money Transfer and Digital Remittance Eco-System

The mobile money clime comprises a chain of firms, people and other operations that connect to manage and promote the payment system delivery (Jenkins, 2008). The digital remittance and mobile money transfer ecosystem consists of MNOs, regulators, commercial and central banks as well as end-users (receivers and senders) and every stakeholder is important in sustaining the ecosystem. The MNOs offer the infrastructure (database, backend server, communication channels) for other stakeholders to associate with. Nevertheless, they have little or no experience in delivering financial services, regulatory oversight, payment risk and legal payment governance (Jenkins, 2008). The

Central Bank offers banking licenses and financial regulatory rules for the MNOs to operate (Rieke et al., 2013). The regulator ensures an environment favourable for innovation, sustainable legislative implementation and financial system stability. The mobile money industry needs telecom and banking regulations (Rieke et al. 2013). The end-users are the beneficiaries of functional and reliable mobile money services.

2.4   Regulatory Issues

To safeguard the end-users of mobile money, new regulations are required to implement compliance with regulatory procedures for the mobile money transfer ecosystem. Hence, there is a need for active role-play of every key player. This regulation are necessary to build the credibility of the mobile money industry and ensure that stakeholders know systematic means of supervising, assessing and identifying business risks (Mauree and Kohli, 2013). Various countries have definite operating criteria for customers using mobile money platforms. This is broadly dependent on the degree of activities of the mobile money operations. Whether or not to subject mobile money operators to the same regulatory guidelines as traditional banks remains a controversial discussion.

2.5   Financial Fraud Anatomy and Prevention

In the fintech ecosystem, cybercrime and credit card fraud are increasing. Credit card fraud gives an unauthorized entity illegal access to a person's credit card details. Sometimes, fraudulent acts are perpetrated without possessing a credit card (Puh and Brkić, 2019). To identify fraud in a financial system, supervised machine learning can be utilized through labelled data. This innovation strengthens and upgrades itself at a slower pace with time. Nevertheless, there is a greater possibility that the advanced framework will be able to recognize fraud corresponding to past fraud patterns. To protect end-users in a financial system, a more advanced security framework should be developed. A self-service should be put in place to safeguard every credit card user from unauthorized access by fraudsters. In addition to a CCTV system, chip authentication technologies can be adopted to mitigate credit card fraud (Button et al., 2009).

2.6   Machine Learning Classifiers

Classification technique is usually used when working with labelled datasets. They are known as the normative systems of handling cases of financial fraud. They are generally applied as non-linear and linear frameworks (Ngai et al., 2011). In previous years, various ML (Machine Learning) classification algorithms such as Decision Trees, Support vector machines and Neural networks have been adopted (Chen et al. 2005; Zaslavsky and Strizhak 2006). Decision Trees are frequently used to score credit risk and detect fraud (Husejinovic, 2020). Husejinovic used naïve Bayes, C4.5 decision tree in a credit card fraud detection experiment. He also used a bagging ensemble ML algorithm to forecast the fraud incidence in a financial transaction. The outcome of the study showed that the C4.5 decision tree gave a higher accuracy of 92.74 percent (Braun et al., 2017). Newer features and pattern information were added to the RF (Random Forest) and LR

(Logistic Regression) models. When these new features were placed side-by-side with the current accrued transaction attributes, both models exhibited a considerable gain in performance. However, the RF ranked higher in performance. RF is used in this study because it can gather non-linear data and offers greater visual outcomes. Although it is scalable, it is susceptible to over-fitting.

The table below shows some relevant works on fraud detection.

Abbr	Full Name	Description
LR	Logistic Regression	Models the probability that one (success) of the two classes of dichotomous events will happen
k-NN	Nearest Neighbour	An instance-based learning method that chooses the k closest instances as it runs
RF	Random Forest	Uses an ensemble of random trees but acts like a decision tree operator

2.7 Theoretical Framework

Routine Activity Theory

Fraudsters are littered on cyberspace plotting to harvest shared personal information and seeking unscrupulous means to hack and defraud users. Fintech users who routinely use the internet are potential victims of fraud due to a greater exposure as a target to likely offenders (Hutchings and Hayes, 2009). Cohen and Felson developed the Routine Activity Theory (RAT) and it is centered around three key elements; a possible offender, a marching target and the lack of a capable guardian (Bottoms and Wile, 1997, p.320). For fraud to happen, there must be a convergence of these three factors. In this context, Users and administrators act as good internet guardians and technology serves as antivirus software, firewalls, and private networks (Yeasmin & Wu, 2021). RAT studies the peculiarity of a victim to study how these victims are affected negatively on the internet (Zhon & Asiama, 2022). The RAT is a primary framework in the study of fraud detection. It is premised on the hypothesis that fraud can be perpetrated by whoever has a means. RAT describes how fraud happens. It offers a holistic view of fraud patterns by evaluating the fraud from various outlooks. RAT forecasts that variations in certain opportunity systems such as technology will heighten the confluence of the three key elements. Some studies on fraud detection have adopted RAT as a theoretical framework while others have not. The RAT aligns with the study aim of creating a fraud detection model utilizing several supervised machine learning techniques to compare different algorithms and attain high detection accuracy and precision.

The assumption that fraudsters are logical when making decisions is a major criticism of the RAT. Fraudsters are not likely to use a similar logical basis as whoever is executing the security initiative The RAT is critiqued for concentration on individual-level components and its constrained applications of structural impacts on frauds. Although RAT focuses on the duties of the three elements in forecasting fraudulent behaviour, it has been faulted for disregarding other societal concerns that aid fraudulent opportunities (Ashely Wellman et al., 2021). Other researchers have emphasized the importance of integrating extra factors beyond the key variables of the theory to offer a broader understanding of fraudulent activities, particularly in the background of cybercrimes (Alex Kigerl, 2021). Scholars who criticized the RAT opined that a more refined perspective that embodies cultural, economic and social factors is important for a comprehensive fraud evaluation. A major limitation of the RAT is in understanding social and behavioural change. While RAT partially describes heterogeneous and intricate behaviours. Furthermore, the emphasis of RAT on patterned activities may disregard the natural interactions between actions and situations that form human patterns (Nikola Banovic et al., 2016).

3 Methodology

The standard ML approach was used in this study. In ML frameworks, the target parameters will be forecast using the tagged class parameters from the financial dataset. The CRISP-DM technique was adopted in this research. It has lingered as the framework for a continuous procedure for data analysis (Schröer et al., 2021). The simulation ecosystem can be categorized into software and hardware divisions. The test was conducted on a 64-bit MacBook Air (with a Processor of 1.1 GHz Quad-Core Intel Core i5, RAM of 8.00 GB) with Python 3.10, Anaconda Navigator 1.10.0, and Jupyter Notebook 6.1.4 installed. The Anaconda Navigator ecosystem adopted the ML technique classifiers Imbalanced-Learn, Pandas, Scikit-Learn, NumPy, Matplotlib and Sea.

3.1 Supervised ML Methodology

The supervised ML algorithm adopts labelled data to guide a classifier to forecast a target category of a new observation, where the result determinants that decide the class observation ought to be reliant on the input parameters (Abdallah et al., 2016). Supervised ML algorithms have been broadly used to recognize fraud in the financial system, particularly when the data to be classified has been assessed by a fraud expert. During the experiment procedure, overfitting happens because of the non-uniform class distribution of non-fraud and fraud incidences. RF is used to figure out regression and classification issues. It offers more precise result forecasts in large data. RF mixes various classifiers to solve intricate cases. RF helps in separating the disadvantages in a DT algorithm (Darwish, 2020).

   The Decision Tree (DT) technique is a useful tool for solving different classification and regression cases. It is the bedrock of the RF algorithm. LR focuses on the connection between the independent variable and the dependent binary variable (Olowookere and Adewale 2020). Unlike DT, LR allows framework overfitting that makes it perform excellently in regression cases, with a low possibility of linear problems. With factual non-linear data, LR is the least approved (Bin Sulaiman et al. 2022). Nevertheless, this study adopted the LR technique to differentiate the outputs of various models.

3.2 Unsupervised ML Methodology

Pattern identification with an unsupervised ML technique is conducted without using target features. This method is favourable for association mining and clustering because every determinant used in the experiment is input (Alloghani et al. 2020). Unsupervised ML methodology is often used in inconsistent recognition methods when a related fraudulent incidence to a suspicious transaction is odd to the main class. (Bolton and Hand 2001). Unsupervised ML algorithms are useful for connecting labels to the data later deployed to conduct supervised learning activities(Hofmann 2001). This means that unsupervised clustering algorithms seek natural collections of data that are not labelled and then label every data value. This study focuses less on unsupervised ML in financial fraud detection (Bolton and Hand 2001) because it is authentic and prioritizes fundamental understanding to clearly explain the newly discovered arrangement of the data. Nevertheless, managing this approach is difficult because for each possible fraud incidence, formulating such regulation needs correct, intricate, and time-demanding programming. The evolving advent of new fraud types demands dynamic rule adaption. More so, the scalability of the system is influenced as a larger dataset needs to be analyzed (Kou et al., 2004).

4   Results and Discussion

Financial data obtained from a 5-year-old remittance startup was used for this study. Hence, the resultant impact for the firm is to adopt the findings and outcomes to strengthen the fraud detection model in the organization. Actual data comprises inconsistent, noisy and missing data majorly connected with their various diverse origins and large size. Data of poor quality often have a considerable impact on the outcome of the study (Han and Kamber, 2006). Therefore, it is important to invest significant time to appropriately make the data ready and suitable for use. The data pre-processing approach utilized on the data includes cleaning, integration, reduction, transformation and visualization.

4.1   Model Design and Development of Fraud Detection System

After the data cleaning, the fraud to non-fraud ratio is estimated at 1:2, with 111,418 frauds and 243,807 transactions. This is an excellent distribution for ML modelling. However, this kind of event can be a signal of a catastrophic incident for any financial firm. The data was divided into test and training sets to develop the ML framework. The training data was maximized to discover the non-obvious sequence in the data and validation or assessment was done using the test dataset. The baseline outcome outputted a test model record for LR: 0.6844397916813, Random Forest: 1.0, KNN: 0.9491730593285945. The outcome of the hyperparameter tuning on the LR revealed zero development as it is the same as the baseline record of 0.6844397916813. The GridSearchCV hyperparameter tuning was conducted with similar grid variables, but the result of the top variables and the model record was unchanged. Even though the baseline model record for the RF framework was 1.0, which is the highest possible model record. Nevertheless, the top variables for the RandomizedSearchCV on the RF are assigned as 'n_estimators':710,'min_samples_split':16,'min_samples_leaf':3,'max_depth':10.

4.2   Performance Metrics and Model Evaluation

Evaluation Measures were applied in assessing the ML approach used in this study. The evaluation measures were chosen premised on its importance to general fraud classification issues. Of all the models used in this study, the RF performed top and it was chosen as the most suitable framework for evaluation. The confusion matrix revealed the TN, FN, TP and FP values. FN represents the amount of fraudulent transactions categorized as legitimate. TP represents the amount of fraud categorized as fraudulent. TN represents the amount of legitimate transactions categorized as legitimate. FP is the amount of legitimate transactions categorized as fraudulent. The confusion matrix showed 48626 True Negatives, 0 False Negatives, 22419 True Positives and 0 False Positives. Precision is the proportion between the amount of positively categorized cases that were correctly categorized and every positively categorized case (Hammouri et al., 2018). The RF framework scored a precision ratio of non-fraud and fraud classification. This shows the most outstanding precision. It also extrapolates a greater degree of credence in the RF model. Recall is the proportion of positive cases that are grouped with those that are overall cases (Ganji and Mannem 2012). The RF framework scored a recall of 1 in non-fraud and fraud classifications, which shows the most outstanding recall. F1-measure is the harmonic average of precision and recall in the test. It is obtained by combining the values of the two measures (Hammouri et al. 2018).

	precision	recall	f1-score	support
0	1.00	1.00	1.00	48626
1	1.00	1.00	1.00	22419
accuracy			1.00	71045
macro avg	1.00	1.00	1.00	71045
weighted avg	1.00	1.00	1.00	71045

Fig. 1. Classification Report

The AUC-ROC is a mechanism used for assessing the classification tasks at various threshold degrees. AUC represents the degree of separability while ROC is the probability curve. It shows how excellent the model can distinguish throughout classes (Narkhede 2018). This study shows that the RF framework is ideal for forecasting fraud incidences as fraud and non-fraud transactions as legitimate transactions (Fig. 1).

5   Recommendations and Conclusion

ML algorithm is useful in recognizing data patterns that distinguish credible users from fraudulent transactions. This is possible by deploying numerous data samples, some of which may infrequently seem to have indistinct linkage to the human assessment (Aleksander 2017). The adoption of ML procedures to forecast financial fraud has generated

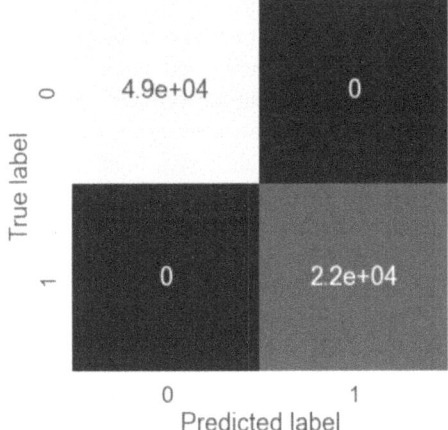

Fig. 2. Confusion Matrix

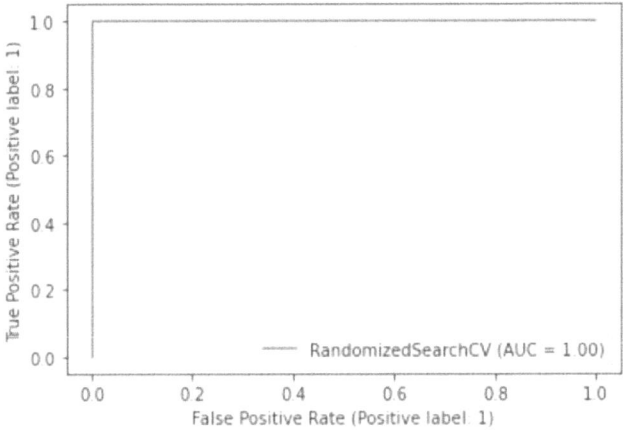

Fig. 3. AUC-ROC Curve

significant outcomes (Zhdanova et al. 2014). An excellent outcome was attained for both unsupervised ML (Bolton and Hand 2001) and supervised ML methods (Kirkos et al. 2007; Khan et al. 2014). The supervised ML technique uses a teacher for a practical learning approach. This predictive framework is equipped with clear data that is labelled or transactions categorized as fraudulent or legitimate to discover associated patterns among the attributes while the unsupervised ML method works with the template of data that are not labelled (Abdallah et al., 2016). Overfitting is common when working with the supervised ML method. Most supervised ML algorithms are unstructured to deal with considerable distinctions in the various instances of different classes(Ganganwar, 2012). Hence, learning from a non-uniform dataset may result in various problems as a result of a classification model. For example, the classifier may assume that the templates are uniformly shared, but in this case, that belief is wrong. The main class is positively displayed by the classification model (Ganguly and Sadaoui, 2017). Many standards can be deployed to measure the computational performance of an algorithm. This entails success potentials, visual mediums and abstract reasoning (Fig. 2).

With actual data offered by a remittance firm headquartered in USA, this study formed the target class by using IBAN counts and EID counts for every transaction. In this study, a binary classification issue was surveyed where a transaction is either legitimate or fraudulent. The study goal is to decide the capabilities of using IBAN count and EID counts for every transaction on the most suitable ML framework for fraud detection. The target class distribution of fraud to legitimate transactions is almost 3:7. This ratio is adequate for an ML framework since the data is uniform. RF, LR and KNN are the 3 ML frameworks used in this study with a precision score between 0.68 and 1.0. The RF performed better than other ML models with a 1.0 score. The RF framework is the most outstanding model in this study following hyperparameter tuning. This study was restricted to categorizing the fraudulent transaction from actual data by the variables decided by the in-house fraud monitoring department of the remittance organization. The variable deployed in the categorization was a wide viewpoint on fraud classification but this is an important phase for the firm to keep innovating a fraud detection model. In subsequent studies, this research can cover the applications of deep learning algorithms to identify fraud in mobile money remittances (Fig. 3).

References

Abdallah, A., Maarof, M.A., Zainal, A.: Fraud detection system: A survey. J. Netw. Comput. Appl. **68**, 90–113 (2016)

Asamoah, D., Takieddine, S., Amedofu, M.: Examining the effect of mobile money transfer (MMT) capabilities on business growth and development impact. Inf. Technol. Dev. **26**(1), 146–161 (2020)

Atanu, D., Jung, K.H., Gill, R.: Mobile Money Opportunities for Mobile Operators. Business & Network Consulting, Huawei Technologies White Paper. 1–22 (2014)

Awanis, A., Lowe, C., Andersson-Manjang, S.K., Lindsey, D.: State of the Industry Report on Mobile Money - GSMA. (2022). https://www.gsma.com/sotir/wp-content/uploads/2022/03/GSMA_State_of_the_Industry_2022_ExecSummary_English.pdf Accessed July 15, 2022

Braun, F., Caelen, O., Smirnov, E.N., Kelk, S., Lebichot, B.: Improving card fraud detection through suspicious pattern discovery. 2017. Springer (2017)

Button, M., Lewis, C., Tapley, J.: A better deal for fraud victims: Research into victims' needs and experiences (2009)

Chris, D.: Alipay ups Africa reach with Ecobank deal. Mobile World Live (2020). https://www.mobileworldlive.com/money/news-money/alipay-ups-africa-reach-with-ecobank-deal Accessed July 18

Darwish, S.M.: An intelligent credit card fraud detection approach based on semantic fusion of two classifiers. Soft. Comput. **24**(2), 1243–1253 (2020)

Finextra: MoneyGram forms money transfer partnership with Airtel. MoneyGram. (2020). https://www.finextra.com/pressarticle/83787/moneygram-forms-money-transfer-partnership-with-airtel#:~:text=The%20partnership%20enables%20Airtel%20Money's,and%20friends%20around%20the%20world. Accessed July 18

Ganganwar, V.: An overview of classification algorithms for imbalanced datasets. International Journal of Emerging Technology and Advanced Engineering **2**(4), 42–47 (2012)

Ganguly, S., Sadaoui, S.: Classification of imbalanced auction fraud data. 2017. Springer (2017)

GSMA: Mobile money: competing with informal channels to accelerate the digitization of remittances (2018)

Guermond, V.: Whose money? Digital remittances, mobile money and fintech in Ghana. Journal of Cultural Economy, 1–16 (2022)

IFAD: Remittances in crisis: response, resilience, recovery. Rome: International Fund for Agricultural Development. (2020). https://gfrid.org/whats-on/the-rctfs-blueprint-for-action-final-report-is-out/ Accessed July 16, 2022

Jenkins, B.: Developing mobile money ecosystems. International Finance Corporation and Harvard Kennedy School, Washington, DC (2008)

Marsland, S.: Machine learning: an algorithmic perspective. Chapman and Hall/CRC (2011)

Mauree, V., Kohli, G.: The Mobile Money Revolution. Part 2: Financial Inclusion Enabler. ITU-T Technology Watch Report. Available online: https://www.itu.int/dms_pub/itu-t/oth/23/01/T23 010000200002PDFE. pdf (accessed on 16 May 2021) (2013)

Ngai, E.W.T., Hu, Y., Wong, Y.H., Chen, Y., Sun, X.: The application of data mining techniques in financial fraud detection: A classification framework and an academic review of literature. Decis. Support Syst. **50**(3), 559–569 (2011)

Ofcom: Ofcom's future approach to mobile markets. UK (2022)

Olowookere, T.A., Adewale, O.S.: A framework for detecting credit card fraud with cost-sensitive meta-learning ensemble approach. Scientific African **8**, e00464 (2020)

Puh, M., Brkić, L.: Detecting credit card fraud using selected machine learning algorithms. 2019. IEEE (2019)

Rieke, R., Zhdanova, M., Repp, J., Giot, R., Gaber, C.: Fraud detection in mobile payments utilizing process behavior analysis. 2013. IEEE (2013)

Schröer, C., Kruse, F., Gómez, J.M.: A Systematic Literature Review on Applying CRISP-DM Process Model. Procedia Computer Science **181**, 526–534 (2021)

Stein, P.: IFC Mobile Money Study. International Finance Corporaton, Washington DC (2011)

Tsys: 2017 TSYS U.S. Consumer Payment Study. TSYS (2017)

West, J., Bhattacharya, M.: Mining financial statement fraud: An analysis of some experimental issues. 2015. IEEE (2015)

Wewege, L., Lee, J., Thomsett, M.C.: Disruptions and digital banking trends. Journal of Applied Finance and Banking **10**(6), 15–56 (2020)

A Cognitive Analysis of CEO Speeches and Their Effects on Stock Markets

Rohan Manro[1], Rui Mao[2(✉)], Liza Dahiya[3], Yu Ma[4], and Erik Cambria[2]

[1] Indian Institute of Technology, Goa, India
rohan.manro.21042@iitgoa.ac.in
[2] Nanyang Technological University, Singapore, Singapore
{rui.mao,cambria}@ntu.edu.sg
[3] Indian Institute of Technology, Bombay, India
lizadahiya@cse.iitb.ac.in
[4] Wuxi Institute of Administration, Wuxi, China
mayu@njust.edu.cn

Abstract. The cognitive state of a CEO can have a great impact on the company's operational results and stock market performance. Conventional cognitive analysis often relies on interviews with cognitive scientists or psychologists, which are not readily scalable for big data applications in finance. In this work, we leverage a novel method to analyze the cognitive states of top-tier managers of 14 well-known companies. We analyze the concept mappings from their speeches and metaphorical expressions over 15 years. We also conduct breakdown analysis for the concept mappings, according to the trends of stock prices. We identify four distinct types of stock market performance and illustrate the featured concept mappings associated with each category. These representative concept mappings reflect the cognitive states of CEOs and provide insights into which cognitive states are most likely to correlate with positive stock market performance.

Keywords: Cognitive Analysis · Concept Mapping · FinTech · Metaphor

1 Introduction

In the modern financial landscape, the role of a CEO in influencing their company's stock price cannot be overstated. Time and again, we have seen how CEO speeches and their perception by the general public can lead to significant fluctuations in stock prices. With a surge in unstructured multimedia data, every statement made by CEOs is carefully analyzed. A popular example is Steve Jobs's iconic announcement of the iPhone in 2007, which revolutionized the smartphone industry and significantly impacted Apple's stock price[1]. His

---

[1] https://www.slashgear.com/how-steve-jobs-iphone-keynote-changed-everything-12706925/.

charismatic presentation and the groundbreaking features of the iPhone captivated investors and consumers alike, leading to a surge in Apple's market value.

Several studies have analyzed CEOs' speeches using various NLP techniques to find correlations with stock prices [1,22,24]. Additionally, CEOs' cognition has been studied in depth using various manual methods to understand their impact on overall firm performance, management style, and CSR efforts [3,12,18]. Despite these extensive studies, there remains a significant gap in understanding the cognitive states of CEOs through their speeches and how these cognitive states correlate with stock market performance. This study aims to fill this gap by leveraging a novel approach: analyzing CEOs' speeches in "Letters to Shareholders" from a cognitive perspective using metaphorical concept mappings. We examine CEOs' speeches through a cognitive lens, investigating their relationship with stock markets by analyzing metaphorical concept mappings. Metaphors, which use language to convey meanings beyond their literal interpretations, illustrate the conceptual mappings between target and source domains. These mapping patterns reveal the unique cognitive perspectives of speakers toward specific concepts. As such, metaphors serve as valuable tools for cognitive analysis, as further detailed in Sect. 3. MetaPro[2] [31] is employed to identify metaphorical expressions from text and parse metaphor interpretation and concept mappings. It includes NLP modules in the tasks of metaphor identification [29], metaphor interpretation [30], and concept mapping generation [15].

In this work, we try to answer the question: What are the representative cognitive patterns of CEOs that result in stock prices having a constant or fluctuating rate of growth or decline? The major findings are as follows: (1) CEOs in consistently stock-price-growing companies emphasize action and optimism. The metaphorical concept mappings also suggest a high valuation of stability and quality, linking them to sustained growth through public confidence. (2) Companies with fluctuated growth rates in stock prices share some cognitive patterns with consistently growing companies. However, presenting achievements without future direction creates uncertainty and contributes to fluctuating stock trends. (3) Predominantly negative cognitive patterns among decision-makers in companies experiencing fluctuating declines. Delayed decision-making and instability, fostering a perception of ineffectiveness. However, occasional positive mappings provide temporary optimism, which creates minor fluctuations in an otherwise declining trend. (4) The absence of forward-looking actions in cognitive patterns leads to a consistent negative market trend, as stakeholders remain skeptical about the company's future. In our analysis, the term "CEO" is used as a synecdoche, representing not merely the individual holding the title but the collective decision-making and strategic intent of the company's senior leadership. As the public face of corporate strategy and performance, the CEO symbolizes the entire top-level management's vision and actions. Decisions attributed to the CEO, especially those articulated in formal communications such as the "Letters to Shareholders", are often shaped collaboratively by the executive team, including the Chairman of the Board, CFO, and other key leaders. Thus, the

---

[2] https://metapro.ruimao.tech/.

term "CEO" in our paper embodies the strategic mindset of the entire executive leadership rather than a single individual acting in isolation.

Regarding the fact that the "Letter to Shareholders" may be written by a secretary or other professional writer, it is important to understand that: Although a secretary or communication professional may assist in drafting the "Letter to Shareholders", such letters are nonetheless a reflection of the core strategic messages intended by the company's top-level decision-makers. These letters undergo rigorous reviews by the CEO and other senior executives to ensure that they accurately convey the company's priorities, vision, and performance narrative. The content is thus an authentic representation of the company's official stance and reflects the collective perspective of its executive leadership. The drafting process involves aligning with the strategic insights and directives provided by top-level decision-makers, meaning that even if the exact wording is developed by someone other than the CEO, the essence of the message represents the will and intention of the company's highest leadership.

Together, we argue that the cognitive patterns analyzed in these shareholder letters genuinely represent the thinking and strategic orientation of the company's executive leadership, regardless of whether the letter is written directly by the CEO or by another member of the team.

2   Related Work

2.1   Cognitive Analysis for CEOs' Speeches

Various studies have explored CEOs' cognition to understand how their mental processes influence organizational outcomes. Calori et al. [3] used in-depth interviews to create cognitive maps, identifying key industry concepts and their interconnections. Kiss et al. [18] demonstrated through surveys and experiments that cognitively flexible CEOs perform more persistent and effortful information searches. Eggers and Kaplan [12] conducted a longitudinal study using content analysis of CEOs' letters to shareholders to examine how CEO attention influences strategic decisions. Li et al. [23] used panel regression analysis to show that CEOs with higher cognitive abilities boost both performance and CSR efforts in SMEs. Fernández-Pérez et al. [14] explored how CEO temporal focus affects organizational ambidexterity, showing that balancing a future and present focus enhances ambidexterity. Analyzing CEOs' speeches has been pivotal for predicting stock prices, with sentiment analysis being the most commonly employed technique in this domain [5,6,8,10,25,34]. Researchers [2,24,38] have analyzed correlations between the sentiment of CEO speech transcripts and stock prices. Baker and Wurgler [1] highlighted the role of investor sentiment in stock market fluctuations. Leitch and Sherif [22] examined the relationship between Twitter sentiment on CEO succession announcements and stock returns.

While these studies provide valuable insights, they often focus on sentiment analysis or vocal cues without delving into the deeper cognitive processes underlying CEOs' communications. Our study goes beyond sentiment analysis by employing metaphorical concept mappings to reveal the cognitive patterns of CEOs and their relationship with stock market performance.

2.2 Application of Conceptual Metaphor Theory

Conceptual metaphors have been widely used in cognitive analysis, particularly in political discourse. Charteris-Black [9] contrasted the use of metaphors in British and American political discourse. Negro [37] analyzed metaphors describing political corruption in the Spanish press. Koller [19, 20] examined metaphors in leadership and management, focusing on war metaphors' impact on perceptions of female leadership. With advancements in computational metaphor processing, researchers have uncovered deep cognitive insights by examining how metaphors map concepts. Han et al. [16] and Mao et al. [26, 33, 35] used computational methods to investigate cognitive patterns among financial analysts, public responses to weather disasters, and mental health indicators, demonstrating the impact of metaphors on perception and decision-making.

Despite progress in metaphor processing, comprehensive cognitive analysis related to CEOs' thinking has not been extensively explored with these advanced methods. Our study leverages MetaPro to fill this gap, providing a scalable, automated approach to uncover metaphorical concept mappings in CEO speeches and letters to shareholders, and their correlation with stock market trends.

3 Preliminaries

3.1 Conceptual Metaphor Theory

Metaphors are a fundamental aspect of our everyday language use. Conceptual Metaphor Theory, as proposed by Lakoff and Johnson [21], suggests that metaphors not only enrich our linguistic expressions but also reflect our cognitive processes by mapping elements from a concrete or familiar source domain onto an abstract or complex target domain. This cognitive mechanism allows us to grasp and communicate intricate ideas more effectively. For instance, in the metaphor "I spent two days learning baking", the concrete notion of MONEY (source domain) is used to describe the abstract notion of TIME (target domain), implicating the idea that TIME IS MONEY[3] and highlighting the value of time through the metaphor of financial expenditure. They argued that without metaphors like "magic, attraction, madness, union, nurturance", our understanding of LOVE would be incomplete.

There are several reasons why metaphors are valuable for studying cognition. First, the distinctive patterns of conceptual mappings between source and target domains can reveal unique cognitive processes among individuals. Individuals may employ different metaphors to explain their unique cognitive perspectives on concepts. These distinct cognitive frameworks are likely shaped by

---

[3] In this work, the mapping relationship between a target and a source concept is represented as the form of "a target concept is a source concept.

personal experiences and cultural influences. Second, examining metaphorical expressions from everyday language provides higher ecological validity compared to traditional interview-based psychological assessments, as it captures how people naturally think and communicate in real-world contexts. Third, the advent of automated tools for metaphor processing facilitates large-scale cognitive studies, making it possible to analyze data from extensive corpora and derive insights that are more representative of broader populations [7].

In summary, metaphors serve as a bridge between the familiar and the abstract, enhancing our ability to understand and articulate complex ideas. They offer a powerful lens through which to study cognitive processes, providing a rich source of data that reflects natural language use and enabling large-scale analysis through modern computational techniques.

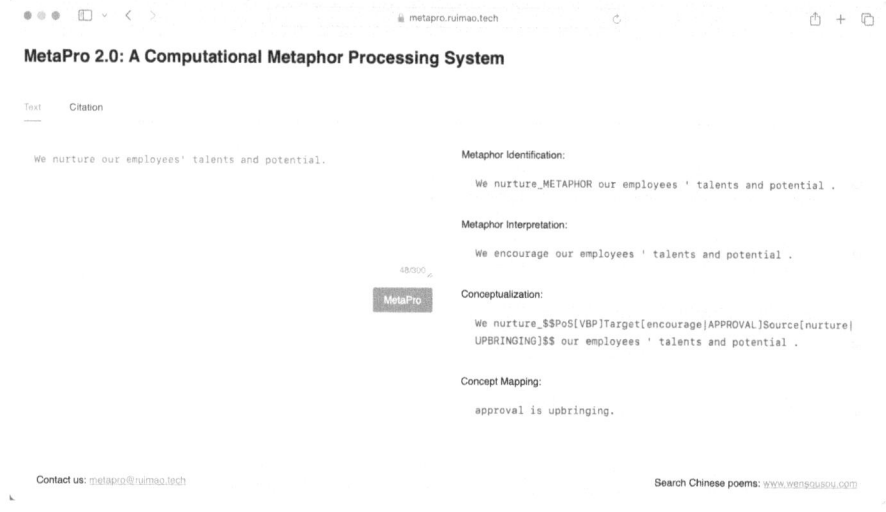

Fig. 1. An example output from MetaPro.

3.2   MetaPro

In this work, MetaPro is used to parse concept mappings from CEO's speech. Then, the findings are delivered upon the statistics of the concept mappings. MetaPro can identify metaphors and generate concept mappings from an open domain. To our knowledge, it is the only tool specifically designed for metaphor processing that can produce end-to-end concept mappings in an open domain [27]. MetaPro consists of 3 technical modules: metaphor identification [29], metaphor interpretation [30], and concept mapping [15]. The latest version of MetaPro is enhanced by a novel pre-training task, termed anomalous language modeling [28]. The detailed algorithms and evaluation can be viewed from these works.

The computational process of MetaPro can be explained from Fig. 1. Given an input, "we nurture our employees' talents and potential", MetaPro first identifies "nurture" is a metaphor. This is achieved by a multitask learning framework [29], which identifies both metaphoricity and Part of Speech (PoS) for each token. The PoS tagging task works as an auxiliary task that enhances the metaphor identification accuracy, meantime, it provides useful knowledge for the following metaphor interpretation task. Next, the metaphor interpretation module paraphrases the identified metaphor, e.g., "nurture" into its literal counterpart "encourage". Such linguistic paraphrasing can enhance the natural language understanding ability for downstream tasks [4, 32]. The paraphrased text serves as the seed word for generating the target concept in the subsequent concept mapping phase. This paraphrase is predicted from a pre-trained language model, which identifies the most probable word to appear at the position of the identified metaphor [30]. The word is one of synonyms or hypernyms of the original metaphor, sharing the same PoS, where the synonyms and hypernyms are obtained from WordNet [13]. Finally, the concept mapping module abstracts the target concept from the paragraph word; the source concept is abstracted from the original metaphor [15]. The target concept represents the domain that is being understood or explained through the metaphor. It is the concept that the metaphor is intended to illuminate. The source concept represents the domain that provides the conceptual framework through which the target domain is understood. It is the familiar domain that is used to make sense of the unfamiliar or abstract target concept. In the example in Fig. 1, the target concept, APPROVAL, and the source concept, UPBRINGING are abstracted from "encourage" and "nurture", respectively, representing the underlying conceptual domains and their mapping relationship, e.g., APPROVAL IS UPBRINGING. The metaphor conveys the speaker's endorsement of the employees' talents and potential by means of supporting their career development.

Table 1. Data Statistics. # denotes "the number of".

		Text
Raw Text	# speeches	222
	# companies/CEOs	14
	Avg. len. of speeches	4,131
MetaPro Output	# concept mappings	40,880
	# unique concept mappings	10,245
	# unique source concepts	1,182
	# unique target concepts	1,016

4   Data

In this study, we analyzed speeches from "Letters to Shareholders" authored by the CEO or Chairman of the Board of Directors, with the objective to gain insights into the cognitive frameworks of top decision-makers. To enhance the depth and breadth of our analysis, we selected companies from diverse sectors, including Investment Banking, Technology, and Energy. We collected letters from 2008 to 2024, covering 14 leading companies in these three sectors (see Table 1). This period was specifically chosen to capture the transition from high sentiment during the 2008 market crash and its aftermath, through a stable period of low sentiment, to another period of high sentiment in 2020. These sentiment transitions provided a compelling timeline for analysis, as demonstrated in the works of Baker and Wurgler [1].

5   Findings

We conducted an analysis of the data to identify metaphors and correlate them with the cognitive processes of CEOs. By employing statistical methods on the results derived from the metaphor analysis, we examined trends in the company's stock prices to categorize the companies into distinct groups, namely companies with consistent growth rates, fluctuating growth rates, fluctuating decline rates, and consistent decline rates. This classification allowed us to systematically explore the relationship between the metaphors used by CEOs and the resulting impacts on company performance, as reflected in stock price movements.

Following the Consistent/Fluctuating Rates of Growth/Decline, we applied a meticulously curated criterion to identify representative target-to-source concept mappings. These mappings, characteristically found within each category, significantly influence public opinion and provide insight into the cognition of decision-makers within these companies. Representative target-to-source concept mappings were identified by normalizing the frequencies of all target-to-source concept mappings and selecting mutually exclusive keys within the first quartile of these normalized frequencies. This ensured that the selected mappings were characteristic of the given category. For the remaining common keys, we employed a threshold to filter out those without significant overlap with other categories. This method allowed us to isolate the key cognitive patterns that distinguish each group, thereby offering a deeper understanding of how executive cognition shapes corporate strategies and public perceptions.

Figure 2a illustrates that decision-makers predominantly focus on taking action in challenging situations and maintaining an optimistic outlook, rather than expressing concern. This cognitive pattern is evident from the warm regions corresponding to the ACTIVITY IS MERCANTILE_ESTABLISHMENT, ACTION IS SITUATION, and IMPORTANCE IS SIZE mappings, while cooler regions are observed in the CONCERN IS STATUS mappings. An additional factor contributing to the

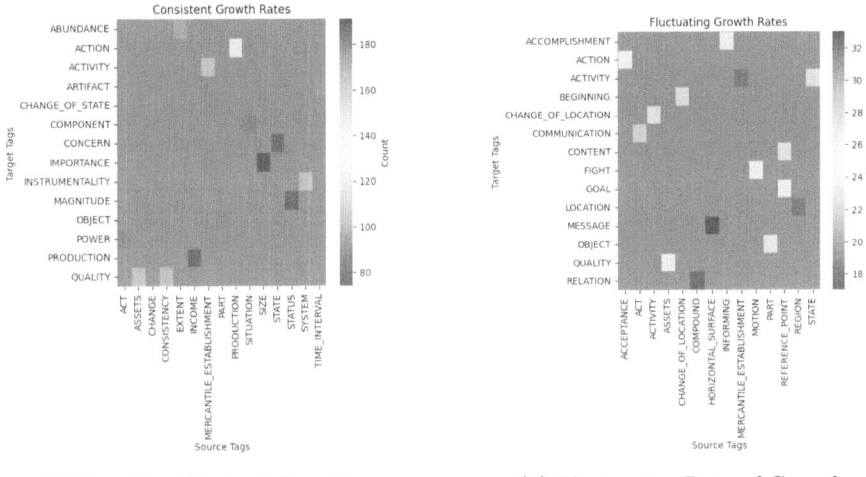

(a) Consistent Rate of Growth.   (b) Fluctuating Rate of Growth.

Fig. 2. Growth categories concept mappings.

positive market trends for companies in this category is the connotation of quality associated with consistency, where quality is perceived as an asset. This cognitive pattern is reflected in the higher frequency of QUALITY IS CONSISTENCY and QUALITY IS ASSETS mappings.

These metaphorical concept mappings suggest that decision-makers in consistently growing companies prioritize stability and value, thereby reinforcing public confidence and contributing to sustained growth. This emphasis often attracts large-scale investors and, due to the herding phenomenon, other investors are inclined to follow [17]. Figure 2b reveals that companies with fluctuating growth rates convey connotations similar to those of companies with consistent ones. However, concept mappings in these companies indicate different nuances. For instance, the FIGHT IS MOTION mapping suggests a delay in conflict resolution, indicating that issues are acknowledged but not promptly addressed, while ACCOMPLISHMENT IS INFORMATION highlights the communication of achievements to the public, yet this information is presented without references to future implications or plans, creating uncertainty about the company's future direction.

These cognitive patterns imply that while the companies project a positive image by acknowledging achievements, the lack of clarity and delayed resolutions may lead to public skepticism regarding long-term stability and growth. A similar instance was found in the study Zhou, Xu and Zhao [40]. This uncertainty is reflected in the fluctuating stock trends, as investors and stakeholders may find it challenging to maintain confidence in the company's sustained success. By understanding these metaphorical mappings, we can better comprehend the underlying factors contributing to the fluctuating growth rates and the resultant market behaviors. Figure 3a shows that many concept mappings in the letters to shareholders, reflecting the cognition of decision-makers in companies within this category, carry negative connotations. Concept mappings like CASUAL_AGENT IS CAPABILITY, FIGHT IS MOTION, and CHANGE_OF_LOCATION IS ACTIVITY depict an image of delayed decision-making, instability, and insufficient seriousness in critical activities.

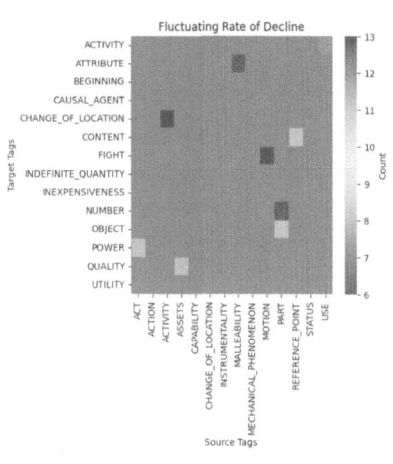

(a) Fluctuating Rate of Decline.

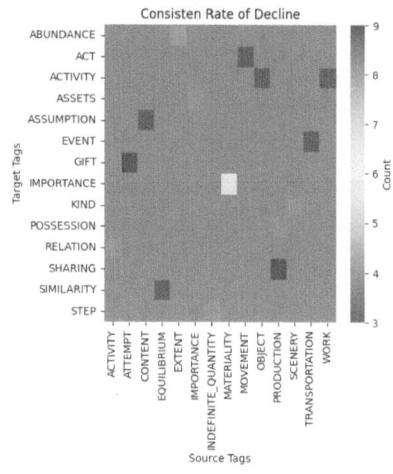

(b) Consistent Rate of Decline.

Fig. 3. Decline categories concept mappings.

These negative cognitive patterns contribute to an overall perception of ineffectiveness and volatility. However, the presence of some redeeming concept mappings, which are typically found in companies with positive growth rates, introduces fluctuations within the otherwise negative market trends. These positive mappings provide intermittent stability and optimism, temporarily alleviating the negative perceptions. Understanding these cognitive patterns helps explain the inconsistent performance and market behavior observed in companies experiencing a fluctuating decrease. The negative connotations trigger loss aversion behavior among investors. However, due to the gambler's fallacy, which arises from certain redeeming qualities, fluctuations are inevitable [11].

In Fig. 3b (the heat map for consistent decrease), although there are no frequent concept mappings that leave a distinctly negative impression, there is a notable and subtle difference from other categories. Mappings such as IMPORTANCE IS MATERIALITY, GIFT IS ATTEMPT, POSSESSION IS PRODUCTION, EVENT IS TRANSPORTATION, and ABUNDANCE IS EXTENT, while carrying positive connotations, fail to imply any significant achievements, conflict resolutions, or future plans [39]. This results in an unclear and often deceptive portrayal of the company's operations and decision-making processes. This lack of clarity and direction is a contributing factor to the consistently negative market trends observed in these companies. The subtle yet significant absence of forward-looking or resolute actions in these metaphorical mappings undermines stakeholder confidence, leading to sustained declines in market performance. Once the company is stuck in this trend, it gets difficult to get out of [36].

6   Conclusion

In this work, we leveraged the state-of-the-art cognitive analysis tool, MetaPro to analyze the cognitive states of CEOs. We link the cognitive patterns to the stock market performance of the companies and deliver insightful findings. We found that decision-makers in companies with consistently growing stock prices focus on action, valuing stability and quality, which boosts public confidence and sustained growth.

In contrast, companies with fluctuating growth rates often fail to provide a clear future direction when presenting achievements, leading to uncertainty and variable stock trends. Companies experiencing declining trends typically show negative cognitive patterns like delayed decision-making and instability, perceived as ineffective, though occasional positive insights cause slight improvements in their overall downward trend. A lack of proactive, forward-looking actions in some companies contributes to persistently negative market perceptions and trends, as stakeholders doubt the company's future prospects.

From this analysis, it emerges that the cognitive orientation of CEOs plays a critical role not only in navigating the current business environment, but also in shaping perceptions of companies' future potential. The ability of CEOs to communicate stability, action, and a coherent forward-looking strategy translates directly into the market trust, which in turn influences the financial trajectory of their organizations. This interaction suggests that enhancing CEOs' cognitive consistency in long-term value-driven thinking may be a powerful lever for improving market performance. Thus, developing executive training programs that emphasize strategic foresight, adaptability, and effective communication of vision may offer tangible benefits to companies seeking to improve their market position and achieve sustained growth.

References

1. Baker, M., Wurgler, J.: Investor sentiment in the stock market. J. Econ. Perspect. **21**(2), 129–151 (2007)
2. Bannier, C.E., Pauls, T., Walter, A.: CEO-speeches and stock returns. In: VfS Annual Conference 2017: Alternative Structures for Money and Banking (2017)
3. Calori, R., Johnson, G., Sarnin, P.: CEOs' cognitive maps and the scope of the organization. Strateg. Manag. J. **15**(6), 437–457 (1994)
4. Cambria, E.: Understanding natural language understanding. Springer, Cham (2024). ISBN 978-3-031-73973-6
5. Cambria, E., Howard, N., Hsu, J., Hussain, A.: Sentic blending: scalable multimodal fusion for continuous interpretation of semantics and Sentics. In: IEEE SSCI, pp. 108–117 (2013)

6. Cambria, E., Mao, R., Chen, M., Wang, Z., Ho, S.B.: Seven pillars for the future of artificial intelligence. IEEE Intell. Syst. **38**(6), 62–69 (2023)

7. Cambria, E., Rajagopal, D., Olsher, D., Das, D.: Big social data analysis. In: Akerkar, R. (ed.) Big Data Computing, chap. 13, pp. 401–414. Chapman and Hall/CRC, Boca Raton (2013)

8. Cambria, E., Zhang, X., Mao, R., Chen, M., Kwok, K.: SenticNet 8: fusing emotion AI and commonsense AI for interpretable, trustworthy, and explainable affective computing. In: Proceedings of HCII (2024)

9. Charteris-Black, J.: Why "an angel rides in the whirlwind and directs the storm": a corpus-based comparative study of metaphor in British and American political discourse. In: Advances in Corpus Linguistics, pp. 133–150. Brill (2004)

10. Chaturvedi, I., Ong, Y.S., Tsang, I., Welsch, R., Cambria, E.: Learning word dependencies in text by means of a deep recurrent belief network. Knowl.-Based Syst. **108**, 144–154 (2016)

11. Cheng, Z.: Psychology analysis of investors from the perspective of behavioral finance. In: 2022 International Conference on Economics, Smart Finance and Contemporary Trade (ESFCT 2022), pp. 727–733 (2022)

12. Eggers, J.P., Kaplan, S.: Cognition and renewal: comparing CEO and organizational effects on incumbent adaptation to technical change. Organ. Sci. **20**(2), 461–477 (2009)

13. Fellbaum, C.: WordNet: An Electronic Lexical Database. Bradford Books, Bradford (1998)

14. Fernández-Pérez, V., García-Morales, V.J., Pullés, D.C.: Entrepreneurial decision-making, external social networks and strategic flexibility: the role of CEOs' cognition. Eur. Manag. J. **34**(3), 296–309 (2016)

15. Ge, M., Mao, R., Cambria, E.: Explainable metaphor identification inspired by conceptual metaphor theory. In: Proceedings of the AAAI Conference on Artificial Intelligence, vol. 36, no. 10, pp. 10681–10689 (2022)

16. Han, S., Mao, R., Cambria, E.: Hierarchical attention network for explainable depression detection on Twitter aided by metaphor concept mappings. In: Proceedings of COLING, pp. 94–104 (2022)

17. Kengatharan, L., Navaneethakrishnan, K.: The influence of behavioral factors in making investment decisions and performance: study on investors of Colombo stock exchange, Sri Lanka. Asian J. Fin. Account. **6**, 1 (2014)

18. Kiss, A.N., Libaers, D., Barr, P.S., Wang, T., Zachary, M.A.: CEO cognitive flexibility, information search, and organizational ambidexterity. Strateg. Manag. J. **41**(12), 2200–2233 (2020)

19. Koller, V.: Businesswomen and war metaphors: 'possessive, jealous and pugnacious'? J. Socioling. **8**(1), 3–22 (2004)

20. Koller, V.: Critical discourse analysis and social cognition: evidence from business media discourse. Discourse Soc. **16**(2), 199–224 (2005)

21. Lakoff, G., Johnson, M.: Metaphors We Live By. University of Chicago Press, Chicago (1980)

22. Leitch, D., Sherif, M.: Twitter mood, CEO succession announcements and stock returns. J. Comput. Sci. **21**, 1–10 (2017)

23. Li, H., Hang, Y., Shah, S.G.M., Akram, A., Ozturk, I.: Demonstrating the impact of cognitive CEO on firms' performance and CSR activity. Front. Psychol. **11** (2020)

24. Loughran, T., McDonald, B.: When is a liability not a liability? Textual analysis, dictionaries, and 10-ks. J. Finance **66**(1), 35–65 (2011)

25. Ma, Y., Mao, R., Lin, Q., Wu, P., Cambria, E.: Quantitative stock portfolio optimization by multi-task learning risk and return. Inf. Fusion **104**, 102165 (2024)
26. Mao, R., Du, K., Ma, Y., Zhu, L., Cambria, E.: Discovering the cognition behind language: financial metaphor analysis with MetaPro. In: Proceedings of IEEE ICDM, pp. 1211–1216 (2023)
27. Mao, R., et al.: A survey on pragmatic processing techniques. Inf. Fusion **114**, 102712 (2025)
28. Mao, R., He, K., Ong, C.B., Liu, Q., Cambria, E.: MetaPro 2.0: computational metaphor processing on the effectiveness of anomalous language modeling. In: Findings of the Association for Computational Linguistics: ACL, pp. 9891–9908. Association for Computational Linguistics, Bangkok, Thailand (2024)
29. Mao, R., Li, X.: Bridging towers of multi-task learning with a gating mechanism for aspect-based sentiment analysis and sequential metaphor identification. In: Proceedings of AAAI, pp. 13534–13542 (2021)
30. Mao, R., Li, X., Ge, M., Cambria, E.: MetaPro: a computational metaphor processing model for text pre-processing. Inf. Fusion **86–87**, 30–43 (2022)
31. Mao, R., Li, X., He, K., Ge, M., Cambria, E.: MetaPro online: a computational metaphor processing online system. In: Proceedings of ACL, pp. 127–135 (2023)
32. Mao, R., Lin, C., Guerin, F.: Word embedding and WordNet based metaphor identification and interpretation. In: Proceedings of the 56th Annual Meeting of the Association for Computational Linguistics (ACL), vol. 1, pp. 1222–1231. Association for Computational Linguistics, Melbourne, Australia (2018)
33. Mao, R., Lin, Q., Liu, Q., Mengaldo, G., Cambria, E.: Understanding public perception towards weather disasters through the lens of metaphor. In: Proceedings of IJCAI (2024)
34. Mao, R., Liu, Q., He, K., Li, W., Cambria, E.: The biases of pre-trained language models: an empirical study on prompt-based sentiment analysis and emotion detection. IEEE Trans. Affect. Comput. **14**(3), 1743–1753 (2023)
35. Mao, R., Zhang, T., Liu, Q., Hussain, A., Cambria, E.: Unveiling diplomatic narratives: analyzing United Nations Security Council debates through metaphorical cognition. In: Proceedings of CogSci (2024)
36. Meta, R.: Behavioral Finance: The Psychology of Investing. Credit Suisse Securities LLC, Finance White Paper, pp. 3–6 (02 2015)
37. Negro, I.: 'corruption is dirt': metaphors for political corruption in the Spanish press. Bull. Hisp. Stud. **92**(3), 213–238 (2015)
38. Qin, Y., Yang, Y.: What you say and how you say it matters: predicting stock volatility using verbal and vocal cues. In: Proceedings of ACL, pp. 390–401 (2019)
39. Valdivia, A., Luzón, V., Cambria, E., Herrera, F.: Consensus vote models for detecting and filtering neutrality in sentiment analysis. Inf. Fusion **44**, 126–135 (2018)
40. Zhou, Z., Xu, K., Zhao, J.: Tales of emotion and stock in China: volatility, causality and prediction. World Wide Web **21**(4), 1093–1116 (2017). https://doi.org/10.1007/s11280-017-0495-4

Does E-Money Mediate the Effect of Fundamental Factors on the Stock Price Index?

Diana Zuhroh$^{(\boxtimes)}$ (iD), Rini Setyowati, Sihwahjoeni, Gaguk Apriyanto (iD), and Abdul Malik

University of Merdeka Malang, Jawa Timur, Indonesia
diana.zuhroh@unmer.ac.id

Abstract. This study aims to examine the role of e-money in mediating the effect of fundamental factors, which are also independent variables on different stock price indexes, i.e., the Composite Index (IHSG), LQ 45, Kompas 100, Jakarta Islamic Index (JII), and Srikehati. The independent variables are money supply (X_1), gross domestic product (X_2), and interest rate (X_3). The research data is secondary data using monthly data from July 2009 to March 2023. The money supply measured from the amount of money circulation, GDP using the amount of gross domestic product, interest rate from BI rate, e-money (Y_1) from the volume of e-money transactions, and stock price index (Y_2) from the closing data of each type of index. Data analysis used a path analysis. The results show that money supply and GDP have a positive effect, while the interest rate has a negative effect on e-money and stock price index for Composite Index, LQ 45, Kompas 100, and Srikehati. While for JII, money supply and interest rates do not effect the stock price index. As a result, e-money also do not mediate the effect of money supply and interest rates on stock price indexes. It is possible that these different result caused by differences in investor characteristics. This research is important to be continued since understanding of investor characteristics will be very useful for investment analysts and policymakers, especially the Financial Services Authority.

Keywords: Money Supply · GDP · Interest Rate · E-Wallet · Stock Price Index

1 Introduction

The development of electronic payment systems in Indonesia has been widely used. Central Bank of Indonesia (BI) reports electronic money transactions in August 2022 grew 43.24% annually (yoy) (indonesia.go.id October 14, 2022). Meanwhile, for 2023, BI projects that the value of e-money transactions will continue to grow 30.84% compared to 2021 (indonesia.go.id April 3, 2023). The practicality of using e-money has a significant positive effect on consumer lifestyles (Foster, Sukono, and Johansyah 2022), (Khatimah, Susanto, and Abdullah 2019). Currently, prospective investors can make transactions on the stock exchange online using the payment platform provided by the Indonesia Stock Exchange (IDX) (Kompas.com, 2021). Based on Financial Services

Authority (OJK) regulations, currently, transaction settlements on the stock exchange can be done by book transfer or cash (OJK 2019). Hollis Chenery's theory in his book Pattern of Development, the economic development of a country is accelerated through 4 factors, namely increasing domestic demand, expanding exports, import substitution and technological change (Behrman 1981). Through the development of electronic money, it is expected to encourage economic growth, especially in the development of the capital market. The capital market is one of the entities that is expected to be part of the digitalization breakthrough that is currently developing. According to OJK 2020 data, there are 89 companies, the majority of which are 40% market platforms, in addition, 80% of retail investors are under 40 years old. So it is very possible that this type of investor has an interest that tends to utilize digital technology in transacting in the capital market (cnbcindonesia.com January 31, 2022). According to a study, in Thailand, stock performance in the capital market will affect economic growth, which can have an impact on the use of e-money (Aimon, Sentosa, and Mahatir 2021). Therefore, digital development and capital markets should have a very close relationship to support each other in creating a better national economy.

Apart from the development of digital technology, other factors that can affect the capital market are fundamental macroeconomic factors. Fundamental macroeconomic factors such as inflation, interest rates, money supply, Gross Domestic Product, and currency exchange rates affect the stock market and stock price index (Hosseini, Ahmad, and Lai 2011), (Naik and Padhi 2012), (Yahya and Hussin 2012), (Baskara and Sulasmiyati 2017), (Camilleri, Scicluna, and Bai 2019), (Assagaf et al. 2019), (Vatansever 2020). Fundamental macroeconomic variables determine the growth prospects of companies on the stock exchange. The long-term relationship of fundamental variables that have a greater influence on the stock price index is the interest rate and inflation (1971 Tap), (Al-Majali and Al-Assaf 2014), (Aimon, Sentosa, and Mahatir 2021), (A. Conrad 2021). Based on the theory and results of previous research, the background for the author to test the influence of fundamental macroeconomic variables on the stock price index through e-money. The variables money supply, Gross Domestic Product, and interest rates are used in measuring fundamental macroeconomic variables, because these three variables have an effect on e-money (Arifin and Oktavilia 2020).

2 Literature Review

2.1 Stock Price Index and Fundamental Factors that Influence It

Fundamental macroeconomic factors are factors that influence changes in the economic conditions of a region, for example interest rates, inflation rates, GDP figures, and the amount of money in circulation (Al-Tamimi, Alwan, and Rahman 2011). Macroeconomic fundamentals can influence how stock prices perform in the capital market (WORLU and OMODERO 2017). Money supply is the amount of money circulating in society, which will affect the level of consumption.

According to Keynes' theory(Harcourt 1994) that the quantity of money demand is determined by the need for money for 3 purposes, namely demand for transactions, demand for precautionary money, demand for money for speculation. Most investment analysts predict the movement and performance of the capital market using money supply forecasts (Shiblee 2011).Gross Domestic Product is an indicator that measures the level of economic growth of a region, and is measured on the basis of constant price GDP (Setiawan 2020). Any growth in GDP will have an impact on increasing domestic demand, both household consumption and investment consumption (Reddy 2012), (Financial Services Authority 2022). *Interest rate*is the percentage of interest rates set by the government, as a form of monetary policy in regulating the stability of the financial system. Based on empirical tests, it states that interest rates are a factor in influencing stock index returns assessed through stock portfolio proxies (Zhou, 1996), (Setiawan 2020).

2.2 E-Money

*Electronic Money*or E-money is a prepaid electronic payment tool, where a certain amount of money is attached to it, which can be refilled and can be used to finance various transactions at merchants. According to Nizar and Hanifah (2021), E-money has several criteria: a). Issued on the basis of the value of the money deposited in advance, b). Stored electronically in a medium such as a server or chip, c) used as a means of payment, and d). The value of electronic money deposited by the holder is managed by the issuer.

E-money appears as a form of transaction instrument that falls into the claim form classification (through a technological infrastructure that is centered on blockchain), (Adrian and Mancini-Griffoli 2021). E-money is growing faster in various countries in the world because it has advantages including user comfort, complementarity, available ubiquity, low cost transactions, trust, and network effect (Van Hove 1999), (Moharana, Sai, and Ramesh 2013), (Aithal 2016), (Adrian and Mancini-Griffoli 2021). The development of technology through the issuance of electronic money as a means of payment encourages increased transactions and public consumption. The capital market began to develop following the development of technology has implemented online transaction automation for stock exchange players, through regulations on non-cash securities settlement.

2.3 Stock Price Index

The stock price index is an indicator that measures the performance of stock prices on the capital market (Hartono 2016). The number of stock indexes on the Indonesian Capital Market (IDX) in 2021 are 37 indexes, consist of 4 sector classifications, namely headline, sector, thematic, factor (Indonesia Stock Exchange 2021). The stock price index used in this study are:

a. Composite Stock Price Index (IHSG)
    It is an index that measures the performance of all stock prices listed on the Indonesia Stock Exchange, both on the Main Board and the Development Board.
b. LQ45
    It is a stock index that measures the performance of 45 stocks that have a high level of liquidity and market capitalization with good company fundamental conditions.

c. Jakarta Islamic Index (JII)

It is a stock index that measures the performance of 30 shares of sharia companies with good financial performance and a good level of company liquidity.

d. KOMPAS100

KOMPAS100 Index is a stock price index that measures 100 stocks of companies with good financial performance and liquidity levels.

e. SRI-KEHATI

It is a stock price index that measures the performance of 25 listed stocks that have good financial performance in encouraging sustainable business, as well as awareness of the environment, social, and good corporate governance (Sustainability and Responsible Investment).

2.4   Relationship Between Variables and Hypothesis Formulation

The Influence of Money Supply on E-Money

The amount of money circulating in society influences consumer interest in carrying out consumption activities (Khatimah, Susanto, and Abdullah 2019). This behavior will affect consumer motivation to transact with e-money with various promotional facilities and ease of use (Lukmanulhakim, Djambak, and Yusuf 2016), (Qin 2017), (Putri and Prasetyo 2020), (Foster, Sukono, and Johansyah, 2022).

H1: Money Supply has an effect on E-money.

The Impact of GDP on E-Money

Long-term changes in national income levels have an impact on fluctuations in transactions using e-money (Arifin and Oktavilia 2020). Good economic conditions will influence the way people use the value of money (Suseco 2016), (Putri and Prasetyo 2020).

H2: Gross Domestic Product has an effect on E-money.

The Influence of Interest Rates on E-Money

The interest rate policy will affect the money flow cycle in society, so it will have an impact on encouraging the use of e-money as a transaction tool (Allen, Mcandrews, and Strahan 2002), (Ichwani and Nisa 2021), (Ramadhani, Yuwono, and Nugroho, 2021).

H3: Interest Rate has an effect on E-money.

The Influence of E-money on Stock Index.

The impact of issuing e-money provides convenience in transactions (Alghifari Mahdi Igamo1 2018). E-money will have an impact on the company's financial performance and influence stock prices in the capital market (Thomas 2005), (Aimon, Sentosa, and Mahatir 2021), (Brimantyo et al. 2021). Several countries that have implemented digital capital market transactions, have revealed that electronic money plays a role in the development and improvement of capital market performance (Neama and Saleh 2020), (Igoni, Ogiri, and Boloupremo, 2021).

H4: E-money has an effect on the Stock Index.

The Influence of Money Supply on Stock Indexes.

Money circulation affects investor stock transactions on the stock exchange. So changes in money supply will have an impact on the stock price index on the capital market (Rogalski and Vinso 1977), (Blanchard 2002), (Baskara and Sulasmiyati, 2017).

H5: Money Supply has an effect on the Stock Index.

The Impact of GDP on Stock Price Index.

When a country's economic condition improves in the long term, it will encourage stock performance in the capital market to also increase. Because the community as potential investors have the financial ability to make investments, so it will also have an impact on stock prices (Shiblee 2011), (Naik and Padhi 2012), (Baskara and Sulasmiyati, 2017).

H6: Gross Domestic Product has an effect on the Stock Index.

The Influence of Interest Rates on Stock Indexes.

The interest rate level will determine the sustainability of the company in its business process. In the long term or short term, the interest rate level can affect stock prices in the capital market (Zhou, 1996), (Camilleri, Scicluna, and Bai 2019), (A. Conrad 2021), (Extract 2023).

H7: Interest Rate has an effect on the Stock Index.

The Influence of Fundamental Factors on Stock Indexes via E-money.

Fundamental macroeconomic factors can influence the volume of e-money transactions (Putri and Prasetyo 2020). In addition, fundamental macroeconomic factors such as the amount of money in circulation, GDP value, and interest rates can affect the stock price index (Rogalski and Vinso 1977), (Triani 2013), (Muid and Raharjo 2013), (Baskara and Sulasmiyati 2017). Through electronic money instruments and easy transaction facilities, these factors influence stock prices (Neama and Saleh 2020) (Igoni, Ogiri, and Boloupremo 2021).

H8-a. Money Supply affects the Stock Price Index through E-money.

H8-b Gross Domestic Product has an effect on the Stock Price Index through E-money.

H8-c Interest rate affects the Stock Price Index through E-money.

3   Research Methods

3.1   Research Variables and Measurement

The variables in this study consist of independent variables, namely money supply (X1), gross domestic product (X2), interest rate (X3), intervening variables, namely (Y1), and dependent variables, namely the Stock Price Index (Y2). The data used in the study are secondary data collected using the documentation method and obtained through the Bank Indonesia website, namely www.bi.go.id, Indonesia Stock Exchange, namely www.idx. co.id. And BPS namely bps.go.id. Meanwhile, the type of data used in this research is quantitative data in the form of:

a. Stock Price Index IHSG, LQ 45, Jakarta Islamic Index (JII), KOMPAS100, SRI-KEHATI closing per month from July 2009 – March 2023
b. Transaction volume using e-money per month from July 2009 – March 2023
c. Money Supply using amount of money in circulation per month from July 2009 – March 2023
d. Gross Domestic Product (GDP) using data per month from May 2009- March 2023
e. Percentage of BI Rate using interest rate per month from July 2009- March 2023

Data analysis using path analysis. Before the hypothesis test, descriptive statistical tests were carried out, classical assumption tests consisting of normality tests, multicollinearity tests, heteroscedasticity tests, and autocorrelation tests (Chandrarin 2017). The following Fig. 1 is a path analysis model formulated in the path analysis statistical model.

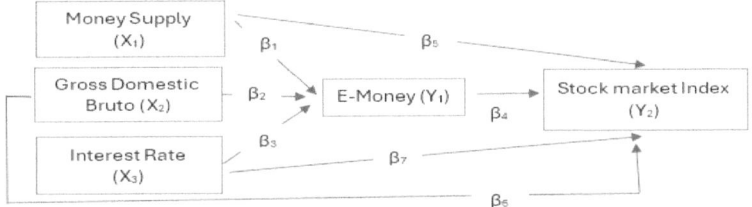

Fig. 1. Path Analysis Statistical Model

4   Research Result

4.1   Descriptive Statistical Test

Statistical tests on 165 data, found 6 outlier data, leaving 159 data consisting of money supply data, GDP, interest rate, IHSG stock price index, LQ45, JII, KOMPAS 100, and SRI-KEHATI from July 2009-March 2023. Table 1 presents the results of descriptive statistics.

Based on the results of descriptive statistical tests, it shows that the minimum value of the five stock indices above occurred in early February 2010. This condition coincided with a decrease in the amount of money supply and was followed by a decrease in GDP in the first quarter of 2010, and was the lowest GDP value throughout the research period. The decrease in the level of per capita income of the community had an impact on the decrease in consumption and utilization of e-money transactions, indirectly impacting the financial condition of the company, and the decrease in stock prices in the capital market. The stock price index with the highest liquidity according to the LQ45, JII,

Table 1. Descriptive Statistical Test Results

	N	Minimum	Maximum	Mean
Money Supply	159	490084.00	2539067.00	1268763.5031
GDP	159	547452.00	996212.00	785031.1321
Interest Rate	159	3.50	7.75	5.65
E-money	159	1914662.00	985462663.00	179337395.9371
IHSG	159	2549.00	7229.00	5177.6226
LQ45	159	496.00	1106.00	842.1509
JII	159	414.00	787.00	616.9119
KOMPAS100	159	611.00	1397.00	1069.5283
SRIKEHATI	159	145.00	440.00	301.4465
Valid N listwise)	159			

Source: Data Processed by Researchers 2023.

and KOMPAS100 stock indices was in January 2018 when GDP, money supply and interest rates were stable before the pandemic hit Indonesia. Meanwhile, for the IHSG and SRI-KEHATI indices, the highest occurred in 2022, after the pandemic recovery and economic conditions began to experience good growth, which had an impact on changes in the stock price index.

To meet the regression requirements, normality test conducted using Normal P-Plot shows that data is normally distributed. Autocorrelation test with Durbin Watson concludes that there is no autocorrelation. Multicollinearity test using tolerance value > 10, VIF < 10 shows that the test results do not have multicollinearity in the regression model. Heteroscedasticity test using Spearman rank shows that the sig. (2 tailed) value > 0.05 means that there are no symptoms of heteroscedasticity, and the classical assumption test on this research variable has been met.

4.2 Hypothesis Testing

Model Accuracy Test (F Test) and R2 Determination Coefficient Test

The following Table 2 presents the results of the model accuracy test and determination test;

Based on the results of the F test, sig < 0,05, it shows that all models formulated in the study are appropriate (fit). Meanwhile, the results of the R^2 test show that the independent variables in this study influence the dependent variables, namely 5 stock price indixes with different values. Among these results, the independent variables have a greater influence on the SRI-KEHATI stock index, namely the stock index that measures Sustainable and Responsible Investment, by 88%.

4.3 T-Test Results

Regression results with 5 dependent variables, namely the stock price index IHSG, LQ45, JII, KOMPAS100, and SRI-KEHATI are presented in the following figure:

Table 2. F and R^2 Test Results

Stock Price Index	F Test		R^2 Test	
	F	Sig	R Square	Adjusted R Square
IHSG	229,599	0.000	0.856	0.853
LQ45	111,969	0.000	0.744	0.737
JII	56,008	0.000	0.593	0.582
KOMPAS100	105,903	0.000	0.733	0.726
SRIKEHATI	306,628	0.000	0.888	0.886

Source: Data Processed by Researchers 2023

The Influence of Money Supply, GDP, and Interest Rate on E-Money

The results of the regression as presented in Fig. 2 shows that the money supply and GDP have a significant positive effect on e-money, while the interest rate variable has a significant negative effect. This means that the higher the money supply and GDP, the higher the use of e-money. Meanwhile, the higher the interest rate, the lower the use of e-money.

The Influence of Money Supply (X1), GDP (X2), and Interest Rate (X3) and E-money (Y1) on the Stock Price Index (Y2)

As explained in the previous section, there are 5 stock price indixes tested in this study including: IHSG, LQ45, JII, KOMPAS100, and SRIKEHATI. The results show that all hypotheses are accepted for the IHSG, LQ45, Kompas 100 and Srikehati. This means that on the four stock indices, money supply, GDP and E-money have a positive effect while the interest rate has a negative effect. Meanwhile, for the JII stock index, only GDP has a positive effect while other variables have no effect.

The Influence of Money Supply (X1), GDP (X2), and Interest Rate (X3) on the Stock Price Index (Y2) via E-money (Y1)

After conducting the analysis of the direct influence between variables, it is necessary to look at the indirect influence. For this purpose, the analysis is carried out by testing the influence of the three independent variables (X1, X2, and X3) on the stock index through the use of E-money. The results of the t-test are presented in the following Table 3:

This result shows that for JII index, e-money does not mediate the influence of money supply, GDP and interest rates on the stock price index. This is because in the JII index, the influence of Y1 on Y2 is not significant (0.849 > 0.05). Unlike the other four indexes, namely IHSG, LQ45, Kompas 100 and Srikehati, the use of e-money is proven to mediate the influence of independent variables on the stock price index.

5 Discussion

This study obtained important findings related to the role of e-money in mediating the influence of fundamental factors on stock prices. Related to the fact that for the JII stock price index with different results from other stock indices, this finding is quite interesting.

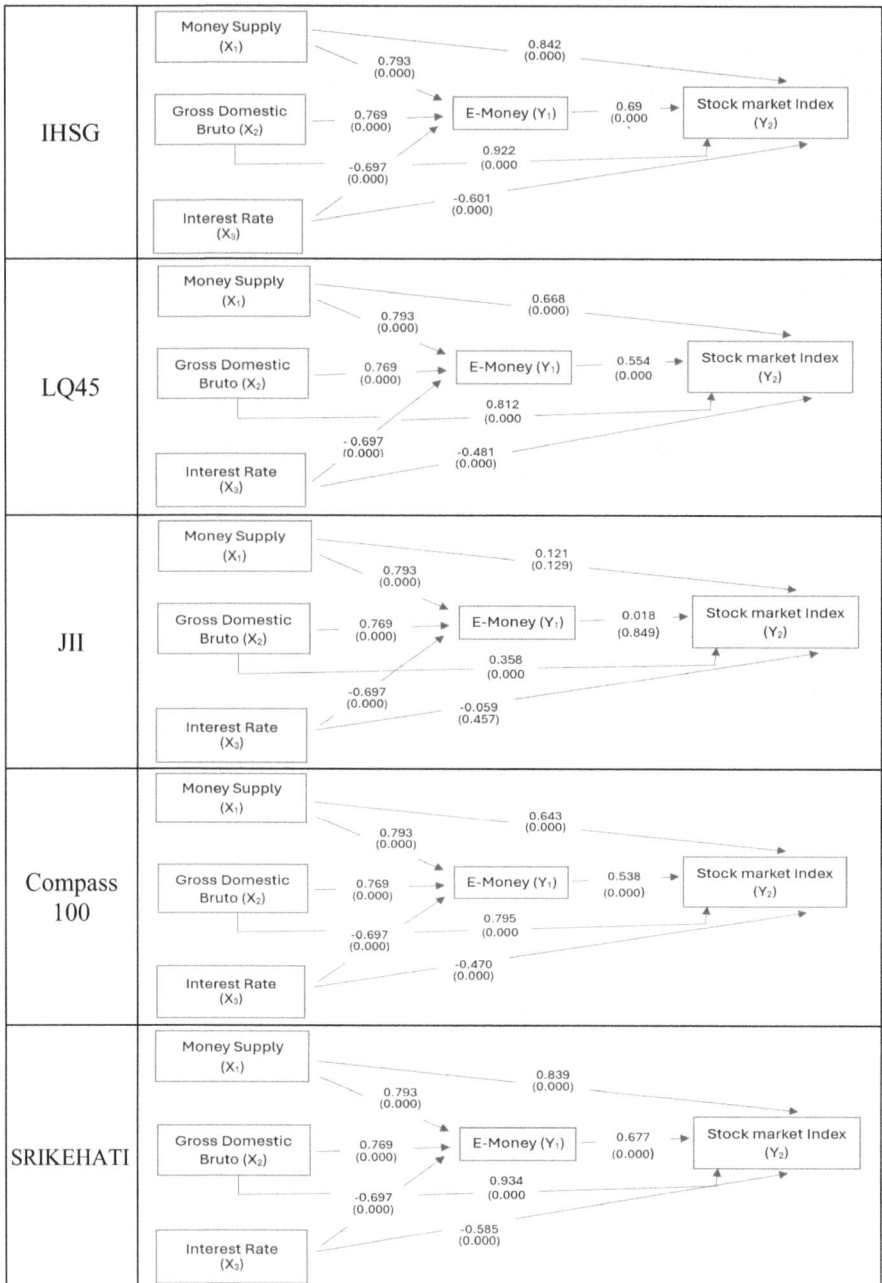

Fig. 2. Regression Coefficient and Significance of 5 Dependent Variables.

Table 3. Test Results Indirect Effects of variables X1, X2, X3 on Y2 through Y1

Index	Relationship Between Var	Indirect Effects	Direct Effects	Total Influence	Hypothesis Rejected/ Accepted
IHSG	X1 → Y1 → Y2	0.547	0.842	1,389	H8-a. Accepted
	X2 → Y1 → Y2	0.531	0.922	1,453	H8-b Accepted
	X3 → Y1 → Y2	-0.481	-0.601	-1,120	H8-c Accepted
LQ-45	X1 → Y1 → Y2	0.440	0.668	1,108	H8-a. Accepted
	X2 → Y1 → Y2	0.426	0.812	1,238	H8-b Accepted
	X3 → Y1 → Y2	-0.386	-0.481	-0.867	H8-c Accepted
JII	X1 → Y1 → Y2	0.014	0.121	0.135 *	H8-a. Rejected
	X2 → Y1 → Y2	0.013	0.358	0.371*	H8-b Rejected
	X3 → Y1 → Y2	-0.012	-0.059	-0.071*	H8-c Rejected
Compass 100	X1 → Y1 → Y2	0.427	0.643	1,070	H8-a. Accepted
	X2 → Y1 → Y2	0.414	0.795	1,209	H8-b Accepted
	X3 → Y1 → Y2	-0.375	-0.470	-0.845	H8-c Accepted
Be careful	X1 → Y1 → Y2	0.537	0.839	1,376	H8-a. Accepted
	X2 → Y1 → Y2	0.520	0.934	1,454	H8-b Accepted
	X3 → Y1 → Y2	-0.472	-0.585	-1,057	H8-c Accepted

* In the JII Stock Price Index, the influence of variable Y1 on Y2 is not significant at the 0.05 level.

Basically, JII is one of three sharia stock indexes in Indonesia. Sharia stocks are stocks where the companies issuing them are companies that have sharia principles and whose operational activities do not violate sharia principles. Therefore, the JII constituents are very limited, consisting of only 30 stocks. Further research is needed, especially regarding the characteristics of investors who invest in these sharia stocks.

The results of this research support the research of Naik and Padhi (2012), Baskara and Sulasmiyati (2017), Vatansever (2020), Setiawan (2020) and Shiblee (2011), that the stock price index is directly and indirectly influenced by fundamental macroeconomic factors. The higher the amount of money supply, the consumer's purchasing power will also increase, then, the use of e-money in society will also increase.

Support previous research (Reddy 2012) show that GDP has a positive effect on the stock price index, both directly and through the e-money variable, because GDP is an indicator of the economic conditions of a country. When the economic conditions indicated by the GDP value incerase, it indicates that the welfare of the community and the economic sector increased, and then affects stock prices. While the interest rate level shows a negative relationship with stock prices which is in line with the research Setiawan2020), because the higher the interest rate, the more a person will tend to use money for operational activities.

Based on the coefficient value, among the 3 fundamental macroeconomic factors, the most dominant factor and the largest direct influence on the stock price index is the GDP. The second factor is money supply, and then interest rate. Meanwhile, e-money has the greatest influence as mediating the interest rate on the stock index.

6 Conclusion

The test results show that the variables of money supply and GDP have a significant positive effect while the interest rate has a significant negative effect on e-money and the stock price index on the IHSG, LQ 45, Kompas 100 and Srikehati. While on the JII index the results obtained are different, where money supply, interest rates, and e-money do not affect the stock price index. The analysis that can be done on the results is that there is a possibility that the difference is caused by differences in investor characteristics. This research is important to continue because understanding investor characteristics will be very useful for investment analysts and policy makers, for example the Financial Services Authority as and the Stock Exchange authority (OJK).

References

Conrad, C.A.: The Effects of Money Supply and Interest Rates on Stock Prices, Evidence from Two Behavioral Experiments. Applied Economics and Finance **8**(2), 33 (2021). https://doi.org/10.11114/aef.v8i2.5173

Adrian, T., Tommaso, M.-G.: The Rise of Digital Money. (2021). https://doi.org/10.1146/annurev-financial-101620

Aimon, H., et al.: E-money and Stock: Empirical Evidence from Indonesia and Thailand. Signifikan: Jurnal Ilmu Ekonomi **10**(1), 139–148 (2021). https://doi.org/10.15408/sjie.v10i1.15380

Aithal, P S.: Realization Ideal Banking Concept Using Ubiquitous Banking. (2016). https://ssrn.com/abstract=2864295

Igamo, A.M., Falianty, T.A.: The Impact of Electronic Money on The Efficiency of The Payment System And The Substitution of Cash In Indonesia. Sriwijaya International Journal Of Dynamic Economics And Business **2**(3), 237–54 (2018)

Allen, F., Mcandrews, J., Strahan, P.: E-Finance: An Introduction. Journal of Financial Services Research. **22**(1), 5–27 (2002)

Al-Majali, A.A., Al-Assaf, G.I.: Long-Run and Short-Run Relationship Between Stock Market Index and Main Macroeconomic Variables Performance in Jordan. Eur. Sci. J. **1010**(1010), 1857–7881 (2014)

Al-Tamimi, H.A., et al.: Factors Affecting Stock Prices in the UAE Financial Markets. Journal of Transnational Management **16**(1), 3–19 (2011). https://doi.org/10.1080/15475778.2011.549441

Arifin, M., Nisa, Q., Oktavilia, S.: Analysis The Use of Electronic Money in Indonesia. Economics Development Analysis Journal **9**(4), 361–373 (2020). https://doi.org/10.15294/edaj.v9i4.39934

Assagaf, A., Murwaningsari, E., Gunawan, J., Mayangsari, S.: The Effect of Macro Economic Variables on Stock Return of Companies That Listed in Stock Exchange: Empirical Evidence from Indonesia. International Journal of Business and Management **14**(8), 108 (2019). https://doi.org/10.5539/ijbm.v14n8p108

BankIndonesia. 2009. *Peraturan Bank Indonesia Nomor :11/12/PBI/2009*

Baskara, Y., Sulasmiyati, S.: Pengaruh Faktor Fundamental Makroekonomi Terhadap Keputusan Investasi Saham Oleh Investor Asing Di Indonesia (Studi Pada Bursa Efek Indonesia (BEI) Periode 2007 â 2014). Jurnal Administrasi Bisnis S1 Universitas Brawijaya **47**(1), 130–139 (2017)

Behrman, J.R.: Review Article On Hollis B. Chenery, Structural Change And Development Policy. (1981)

Blanchard, O.: Output, The Stock Market, and Interest Rates. (2002). https://www.researchgate.net/publication/4723102

Brimantyo, H., et al.: The Role of E-Money on Business Growth in Indonesia. Jurnal Bisnis Dan Manajemen **8**(2), 405–411 (2021). https://doi.org/10.26905/jbm.v8i2.7091

Camilleri, S.J., et al.: Do Stock Markets Lead or Lag Macroeconomic Variables? Evidence from Select European Countries. North American Journal of Economics and Finance **48**, 170–186 (2019). https://doi.org/10.1016/j.najef.2019.01.019

Chandrarin, G.: Metode Riset Akuntansi Pendekatan Kuantitatif. In: Metode Riset Akuntansi Pendekatan Kuantitatif, Puji , P. (eds.) Lestari, pp. 102–106. Malang, Salemba Empat (2017)

Diana, Fatatik Noer, and Widita Kurniasari. 2021. "Buletin Ekonomika Pembangunan Buletin Ekonomika Pembangunan" 2 (2): 116–33

Foster, B., et al.: Analysis of the Effect of Financial Literacy, Practicality and Consumer Lifestyle on the Use of Chip-based Electronic Money Using Sem. Sustainability (Switzerland) **14**(1). (2022). https://doi.org/10.3390/su14010032

Harcourt, G.C.: Kahn and Keynes and the Making of The General Theory.Cambridge Journal of Economics. **18** (1994). https://doi.org/10.1093/oxfordjournals.cje.a035256

Hartono, J.: Teori Portofolio Dan Analisis Investasi, 11th edn. BPFE-Yogyakarta, Yogyakarta (2016)

Hosseini, S.M., Ahmad, Z., Lai, Y.W.: The Role of Macroeconomic Variables on Stock Market Index in China and India. International Journal of Economics and Finance **3**(6) (2011). https://doi.org/10.5539/ijef.v3n6p233

Hove, L.V.: Electronic Money and the Network Externalities Theory: Lessons for Real Life . Netnomics. Vol. 1 (1999)

Ichwani, T., Nisa, C.: Determinants of BI 7 – Days Reverse Repo Rate, GDP, and Exchange Rate Against the Money Supply. Journal of Business, Management, and Accounting **3**(2), 92–99 (2021)

Igoni, S., et al.: The Link between Electronic Transactions and Stock Market Performance in the Nigerian Financial Ecosystem. International Journal of Advanced Engineering Research and Science (IJAERS) **8**(1), 2456–1908 (2021). https://doi.org/10.22161/ijaers

Indonesia Stock Exchange: IDX Stock Index Handbook v1.2. (2021)

Keran, M.W.: Expectations , Money , and the Stock Market. Federal Reserve Bank of St.Louis, pp. 16–31 (1971)

Khatimah, H., Susanto, P., Abdullah, N.L.: Hedonic Motivation and Social Influence on Behavioral Intention of E-Money: The Role of Payment Habit as a Mediator. International Journal of Entrepreneurship **23**(1), 1–9 (2019)

Lukmanulhakim, M., Djambak, S., Yusuf, K.: Pengaruh Transaksi Non Tunai Terhadap Velositas Uang Di Indonesia. Jurnal Ekonomi Pembangunan **14**(1), 41–46 (2016)

Moharana, S.S., et al.: Ubiquitous Virtual Currency. (2013). www.iosrjournals.org

Muid, D., Raharjo, D.: Analisis Pengaruh Faktor-Faktor Fundamental Rasio Keuangan Ter-hadap Perubahan Harga Saham Yang Terdaftar Di Indeks LQ45 Selama Periode 2007–2011. Diponegoro Journal of Accounting **2**(2), 444–454 (2013)

Naik, P.K., Padhi, P.: Interaction of Macroeconomic Factors and Stock Market Index: Empirical Evidence from Indian Data. SSRN Electron. J. (2012). https://doi.org/10.2139/ssrn.2150208

Neama, N.H., Mahdi Saleh, H.: Evaluation of Electronic Trading and Central Depositery. Iraq Stock Exchange for the Perios (2008–2018) **12**(2), 22–33 (2020). https://doi.org/10.28936/jmr acpc12.2.2020

Nizar, M.A., Hanifah, A.: Program Penjaminan Uang Elektronik (E-Money). Fiskal.Kemenkeu.Go.Id, (2021)

OJK: Peraturan Otoritas Jasa Keuangan Republik Indonesia Nomor 22/PJOK.04/2019 Tentang Transaksi Efek. PJOK Tentang Transaksi Efek, pp. 1–29 (2019)

Keuangan, O.J.: CAPITAL MARKET FACT BOOK. Indonesia (2022)

Putri, C.A., Eko Prasetyo, P.: Money Supply, Counterfeit Money, and Economic Growth Effect to E-Money Transaction. Efficient Indonesian Journal of Development Economics **3**(1), 634–649 (2020). https://doi.org/10.15294/efficient.v3i1.35951

Qin, R.: The Impact of Money Supply and Electronic Money: Empirical Evidence from Central Bank in China. BMC Public Health **5**(1), 1–8 (2017)

Ramadhani, W., et al.: Pengaruh Pembayaran Non Tunai Dan Tingkat Suku Bunga Kebijakan Terhadap Sistem Pembayaran Di Indonesia. (2021). https://doi.org/10.14414/jbb.v11i1.2591

Reddy, D V Lokeswar. 2012. "Impact Of Inflation And GDP On Stock Market Returns In India." *International Journal of Advanced Research in Management and Social Sciences* 1 (6): 120–36. www.garph.co.uk

Rizal Ramadhani, W., et al.: Analysis of The Effect of Exchange Rates, E-Money and Interest Rates on the Amount of Money Supply and Its Implications on the Inflation Level in Indonesia 2012–2017, **2**(1), 1–17 (2019)

Rogalski, R.J., Vinso, J.D.: Stock Returns, Money Supply and the Direction of Causality. J. Financ. **32**(4), 1017–1030 (1977). https://doi.org/10.1111/j.1540-6261.1977.tb03306.x

Sari, M.: The Effect Of Money Supply And Interest Rate On Stock Price. Central European Management Journal **31**, 233–40 (2023). https://doi.org/10.57030/23364890.cemj.31.1.24

Setiawan, S.A.: Does Macroeconomic Condition Matter for Stock Market? Evidence of Indonesia Stock Market Performance for 21 Years. The Indonesian Journal of Development Planning. **4** (2020)

Shiblee, L.S.: The Impact of Inflation, GDP, Unemployment, and Money Supply On Stock Prices. SSRN Electronic Journal (2011). https://doi.org/10.2139/ssrn.1529254

Suseco, T.: Effect of E-Money to Economic Performance (A Comparative Study of Selected Countries). In: The 2016 International Conference of Management Sciences, pp. 9–12. (2016)

Thomas, B.N.M.I.: New Stock Market Model. SSRN Electron. J. **1**, 12 (2005). https://doi.org/10.2139/ssrn.403

Triani, L.F.: Faktor-Faktor Yang Mempengaruhi Perubahan Indeks Harga Saham Di Jakarta Islamic Index Selama Tahun 2011. Jurnal Organisasi Dan Manajemen **9**(2), 162–178 (2013)

Vatansever, H.: Determining Impacts on Non-Performing Loan Ratio in Romania. Review of International Comparative Management **20**(2), 119–129 (2020). https://doi.org/10.24818/rmci.2019.2.155

Worlu, C.N., Omodero, C.O.: A Comparative Analysis of Macroeconomic Variables and Stock Market Performances in Africa (2000–2015). International Journal of Academic Research in Accounting, Finance and Management Sciences **7**(4), 3436 (2017). https://doi.org/10.6007/ijarafms/v7-i4/3436

Yahya, M., Hussin, M.: Macroeconomic Variables and Malaysian Islamic Stock Market: A Time Series Analysis. Journal of Business Studies Quarterly **3**(4), 1–13 (2012)

Zandi, M., Sophia, K., Virendra, S., Paul, M.: The Impact of Electronic Financial Payments on Economic Growth. SSRN Electronic Journal, 1–31 (2016)

Zhou, C.: Stock Market Fluctuations and the Term Structure. Finance and Economics Discussion Series **1996**(3), 1 (1996)

Multi-Attribute Decision Making (MADM) Model Based on Bayesian Neural Network Classification with Comparative Scoring Quantification Method Can Solve Intangible Value Valuation Issues in Benefit-Cost Analysis (BCA)

Zhaojie Wang$^{(\boxtimes)}$ (ID) and Hoi-Hei Wang

Imperial College, London SW7 2AZ, UK
`zhaojie.wang124@imperial.ac.uk`

Abstract. This paper is Methodology research of Benefit-Cost Analysis (BCA) under machine learning-driven Classification Bayesian Neural Network. BCA is currently the predominant decision-making tool for corporate capital investment. Due to its inherent single-metric nature, BCA lacks complete quantifiability and responsiveness in terms of intangible benefits and ethics. In this paper the author presents how advanced AI scoring algorithm enables Multi-Attribute Decision Making (MADM) approach to dissolve monetization failure issue stemming from ethical concerns in Benefit-Cost Analysis (BCA). In the current context of increasing emphasis on Corporate Social Responsibility (CSR), ethical considerations have become vital for businesses. Therefore, it is advisable to utilize machine learning leveraging the business big data to train more sophisticated and stable multi-dimensional decision-making models, then integrate them into the framework of BCA, thereby evolving the BCA methodology to overcome its drawbacks. The three main problems associated with BCA discussed are: (1) intangible value is difficult to monetize, and (2) human life-related benefits should not be monetized due to the ethical principle. And (3) the normative and distributional problem of BCA that cannot take into account both the efficiency criterion and the equity criterion in one decision-making process. For the three problems of BCA, author introduces a Bayesian Neural Network Scoring algorithm based Analytic Hierarchy Process (AHP) to achieve Multi-Attribute Decision Making (MADM) and explains why it can solve the above three problems that are difficult for BCA to handle solely. Furthermore, arguing the limitations of such AHP-MADM methodology and puts forward the improvement methods including the Expected Utility Theory [1], and VIKOR method based on "Prospect Theory" [2, 3] which allow each organization to have its own unique decision preferences.

Keywords: Benefit-Cost Analysis · Benefit Valuation · Bayesian Neural Networks · Multi-Attribute Decision Making (MADM) · AHP · Fuzzy · VIKOR · Utility Theory

K.-W. Huang et al. (Eds.): ICFT 2024, CCIS 2437, pp. 45–55, 2025.
https://doi.org/10.1007/978-981-96-3811-6_5

1 Introduction

Multi-Attribute Decision Making (MADM) is a decision-making framework that emerged in the 1980S and has been systematically described by scholars (Hwang and Yoon 1981). It and Multi- Objective Decision Making (MODM) were collectively referred to as Multi-Criteria Decision Making by scholars at that time [4]. The biggest difference between MADM and MODM is that multi-attribute decision making is to rank a finite number of alternatives by analyzing a finite number of attributes, that is, MADM is a discrete comprehensive evaluation method. In contrast, MODM's alternatives are infinite and continuous. Thus, MADM is more suitable for the investment alternatives evaluation in the real world and more like the BCA method. The relationship between MADM and BCA lies in that the two decision-making methods can share the same set of decision-making reference information (Benefits information and Cost information). And in terms of methodology, MADM is better than BCA at dealing with decision-making under fuzzy information and MADM can consider more decision factors(attributes) than BCA. The downside of MADM is that, like Distributionally Weighted BCA, it is bound to involve subjective or objective weighting, and is therefore not considered a mainstream approach for investment evaluation. In this paper, I simulate a investment alternative decision-making by conduct a Analytic Hierarchy Process (AHP) method MADM to showcase how MADM can solve three problems that BCA is difficult to deal with: 1) the monetization of intangible value, 2) the problem that human life should not be monetized due to moral considerations, and 3) the inability to consider both efficiency and equity at the same time.

At the beginning, Sect. 1 discusses the three problems of BCA in investment evaluation practice through a investment Alternative Decision-Making scenario. I point out that the BCA's calculation methodology is inherently unjust when dealing with benefits related to human lives. And I argue why, in theory, the methodology of MADM based on AHP is ethical. Then in Sect. 2, I adopt MADM based on AHP method to conduct a simulated decision-making analysis on a case scenario to demonstrate AHP-MADM can solve the moral problem in practice. To be clear since it is a simulation analysis, the data is not collected from the real world, but this does not affect the validity of the demonstration, because the aim of the case scenario is to discuss the methodology paradigm which nothing to do with the specific data. In the following Sect. 3, I reflect on the defects of AHP-MADM analysis in Sect. 2 and propose improvements. Then in Sect. 4 I practice an improvement and use that as an instance to show how AHP-MADM can consider both efficiency and equity at same time in the decision-making process.

2 Methodological and Application Scenario Analysis

2.1 Three Problems in Practice of BCA

Suppose there is an investment with four alternative implementation options as shown in Fig. 1. And investment analysts have identified three kinds of benefits will be yielded by the investing (each alternative can offer all three benefits, but to varying amounts): Benefit 1 is already monetized, Benefit 2 is "intangible value" that is difficult to monetize and Benefit 3 is "reduced mortality risks" that is controversial to be monetized due to ethical issues. And this investment comes with two kinds of costs, the Cost 1is monetized cost, and the Cost 2 is the cost that is difficult to monetize, such as normative or distributional problems, negative effects on religion and culture. Analysts are now asked to decide which Alternative to use to invest. If we adopt the Benefit-Cost Analysis (BCA) method at this time, we must make efforts to obtain the shadow prices of Benefit 2 and Benefit 3, while avoiding the consideration of Cost 2. There are two problems that need to be considered. First, whether intangible value can be monetized accurately, and second, how normative issues should be handled. For the first question, how to monetize "intangible value". BCA is relatively easy to handle via using Benefit Valuation method. The most common method can be Stated Preference Methods to obtain monetization (utility) of intangible value [5].

Table 1. Comprehensive Matrix of a Investment's Alternatives & Attributes

	Attribute 1 (Benefit 1)	Attribute 2 (Benefit 2)	Attribute 3 (Benefit 3)	Attribute 4 (Cost 1)	Attribute 5 (Cost 2)
Alternative 1	a1	b1	c1	d1	e1
Alternative 2	a2	b2	c2	d2	e2
Alternative 3	a3	b3	c3	d3	e3
Alternative 4	a4	b4	c4	d4	e4

Specifically, if we want to identify people's aesthetic appreciation experience brought by public art exhibitions, we can design questionnaires and recruit respondents to answer the survey then gain people's WTP distribution. But on the second question, there is a big debate. The problem is not technical, but moral. For Benefit 3 "reduced mortality risks", the difference between it and Benefit 2 "intangible value" is that the difficulty to monetize not because of its operational complexity but because it should not be monetized. Even if we do not turn it into a shadow price that takes away the concept of "money" and turn it into "utility" or using Value of Statistical Life (VSL), which is still morally weak. Because according to the BCA process or the Kaldor-Hicks standard, utility of human life will ultimately need to subtract some money-related cost, this addition and subtraction operation itself is immoral. In other words, as long as the BCA method insists on adding and subtracting to gain the final 'Net benefit', it cannot escape the moral condemnation of the treat human life as utility or money. For this issue, Multi-Attribute Decision Making (MADM) framework can help by adopting the Analytic Hierarchy Process (AHP) decision-making method, which avoids comparing different attributes of the same Alternative, but compares the performance of different alternatives on the same attribute. In more detail, the traditional decision-making approach is flat that needs to compare different attributes of a project in the attributes layer (the only layer). The AHP has multiple layers (normally two layers, attributes layer and alternative layer), which allow us to compare the performance of different alternatives in saving human life (one attribute) at alternative layer. Note that here it purely compares which alternative can save more human life, not involves other attributes. It is true that comparing the number of human lives saved by different alternatives may raise deeper moral dilemmas, such as the "trolley problem" of whether the quantity and quality of lives can be compared, but at least it is better than directly comparing lives to other money-related attributes. The second step is that after the Alternative layer comparison, we still need to put the mortality attribute in the Attributes layer to compare with other attributes. But importantly, we need to aware that before entering the Attributes layer and comparing each attribute, we already have alternative ranking results based solely on mortality attribute comparison. Moreover, the most critical thing is that what is compared in the attributes layer is not the value of each attribute, but the weight of each attribute. This weight assignment can be subjective, it can be scored by experts or voted by people, it can be done by any righteous means. For example, a decision maker who believes that saving human lives is far more important than saving public health budget then she may give 99% weight to the former and 1% weight to the latter. Therefore, the process of assigning weight on the attribute layer is not only free of ethical shortcomings, but practices ethics in decision-making. Next, I will present a simulation application senario that uses the AHP-MADM to deal with multiple benefits evaluation problem (Table 1).

2.2 Application of AHP Multi-Attribute Decision-Making

Consider this business decision making scenario that a company consider launch health care plan or a government organization consider initiate health intervention as Fig. 1.

Standard Benefits from a health intervention	
B1 = the intangible value associated with "feeling well"	Net Benefits:
B2 = reduced mortality risks	B1+B2+B3+B4-C, C=cost of the intervention
B3= increased work productivity	B1 usually not monetized (due to difficulty)
B4=health care cost savings	B2 usually not monetized (due to ethical concerns)

Fig. 1. Cost-Utility Analysis in Health Field

To solve the issue of B1 and B2 cannot be monetized, we use neither Cost-efficiency analysis nor Benefit-cost analysis but adopt the multi- attribute decision-making framework with AHP method as shown in Fig. 2.

As shown above, this is a classic three-layer AHP. Blue is the Goal layer, yellow is the Attributes layer, and red is the Alternatives layer. Here I assume that there are three alternatives to the health plan. The first step is to employ Bayesian classification algorithm that uses big data to classify specific situations into the maximum possible score interval to score each attribute (Benefit i). Our assumption here is that the distribution of each rating is Gaussian and that the ratings are independent of each other [6, 7]. Here rather than traditional expert scoring method, I employed Bayesian neural network classification method to achieve quantification. The advantage of this approach is that while Bayesian estimates are based on subjective assumptions, by fine parameter calibration using machine learning, we can incorporate observed past business case experiences big data into our decision-making process, making the results more generalizable and stable.

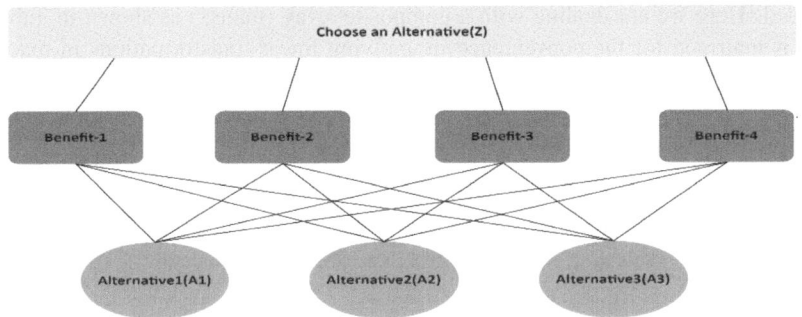

Fig. 2. AHP Layer Structure of MADM (Bayesian Neural Networks)

Table 2. Quantization of Comparison between Attributes

Compare Benefit i to Benefit j	Quantization Value
Equal	1
A little better	3
Better	5
Much better	7
Extremely better	9
The middle of the two judgments	2,4,6,8
Reciprocal	aij = 1/aji

The method of scoring is to compare the attributes (benefits) in pairs. The criteria for comparison are shown in Table 2.

Note, as explained earlier, that this comparison does not need to transform the two attributes' value into a same unit of measure, but rather a direct comparison of preferences. For example, if "Benefit 2 reduced mortality risks" extremely better (more important) than "Benefit 3 increased work productivity", then the comparison score for Benefit 2 on Benefit 3 will be 9, and comparison score for Benefit 3 on Benefit 2 will be 1/9. There is nothing morally wrong with the argument and comparison that "human life is more important than productivity". Furthermore, the scoring judgment was created by machine and derived from a Gaussian distribution, rather than a specific opinion of any individual. Thus, this Bayesian neural network classification with comparative scoring quantification method can avoid the ethical issues involved in the quantification method of monetization

The next step is to perform a Principal Component Analysis (PCA) dimensionality reduction driven by machine learning algorithms to enable our MADM to rank Alternatives. In our specific context, we have four attributes (benefits), but in other applications, there could be tens or even hundreds of attributes. Therefore, dimensionality reduction is needed.[1] Here we are dealing with a composite array (matrix) as shown in Table 3, which is matrixed for the convenience in applying linear transformations in machine learning.

$$C = \frac{1}{n}(A - \mu)^{T}((A - \mu))$$ (1)

Table 3. Composite Matrix [normalization by column]

Z	Benefit 1	Benefit 2	Benefit 3	Benefit 4
Benefit 1	0.12	0.13	0.13	0.09
Benefit 2	0.47	0.53	0.50	0.57
Benefit 3	0.06	0.07	0.06	0.06
Benefit 4	0.35	0.27	0.31	0.28

For preprocessing the composite matrix, here simply adopted arithmetic average method [8]. Then find the average of each row of the normalized matrix to gain the weight value of each 'Benefit' as Table 4.

Table 4. Weight Value of each 'Benefit

Z	Benefit 1	Benefit 2	Benefit 3	Benefit 4	W
Benefit 1	0.12	0.13	0.13	0.09	**0.1176**
Benefit 2	0.47	0.53	0.50	0.57	**0.5175**
Benefit 3	0.06	0.07	0.06	0.06	**0.0611**
Benefit 4	0.35	0.27	0.31	0.28	**0.3038**

The following critical step is consistency checking. Because we used AI learning algorithm to assign a score to each 'Benefit,' we needed to avoid inconsistencies. The basic assumption of the AI algorithms is that each scoring is an independent event and has no memory. This can easily lead to inconsistencies. Here we should not allow: A > B, B > C, C > A. When we do consistency checking we can adopt matrix eigenvalues. According to the definition let us set λ is the eigenvalues of a matrix. Then there exist $MV = M\lambda$ where M is a matrix and V is the eigenvector of this matrix. From this relation we can calculate the maximum eigenvalue of the composite matrix.

$$\lambda_{max} = \sum_{i=1}^{n} \frac{[AW]_i}{nW_i}$$ (2)

---

[1] PCA algorithm is an unsupervised learning for feature extraction. Using formula (1) to get Covariance Matrix of Composite Matrix like Table 3. Then solve det[C-λI] can get Eigenvalue & Eigenvector for PCA and Consistency checking.

Using above formula (2) to gain the maximum eigenvalue of this comprehensive matrix. Here it is $\lambda max = 4.015$. And at this point we need to calculate three important values: 1. **Coincidence Index** $= (\lambda - n)/(n\text{-}1) = (4.015\text{--}4)/(4\text{--}1) = 0.0052$, 2.**Random Index** $= 0.88$ find in the RI table below in Table 5. And 3. **Consistency Ratio** = Coincidence Index / Random Index $= 0.0059$. Here CR $= 0.0059$, the value is much less than 0.1, so we can conclude that the comprehensive matrix passes the consistency checking, that is, there is no inconsistency problem with machine algorithm scoring.

Table 5. RI Reference Table [9]

n	3	4	5	6	7	8	9
RI	0.524	**0.881**	1.11	1.24	1.34	1.40	1.45

Source: Alonso, J. A., ve Lamata, M. T. (2006) Consistency in the analytic hierarchy process: a new approach. International Journal of Uncertainty, Fuzziness and Knowledge-Based Systems, 14(4): 449

At this point, we processed the index (attributes) layer and obtained the weight value of each 'Benefit' based on AI scoring. Next one should be calculating the weight value of each investment alternative for a certain 'Benefit'. The overall process is similar to calculating attributes weights. And the results of Benefit 1 are shown in Table 6.

Table 6. Benefit 1 Composite Matrix and its Weights

Benefit 1 "Feling well"	Alternative 1	Alternative 2	Alternative 3	WB1
Alternative 1	0.14	0.14	0.14	**0.14**
Alternative 2	0.57	0.57	0.57	**0.57**
Alternative 3	0.28	0.28	0.28	**0.28**

Observing the values in each row of the Table 6. Are the same and each row value is the weight value, which is due to the fact that using objective statistics data of each Alternative in the pairwise comparison of alternatives. This not only shows the difference between subjective and objective value assignment methods, but also proves the rationality of calculating arithmetic weights directly without pairwise comparison when an attribute's objective data is available. Using the same process as for Benefit 1 to obtain composite matrices of the remaining attributes relative to the three alternatives see Table 7 below.

After all the weight values have been calculated, they can be combined into a weight values matrix. Through the weight matrix, it can more easily perform weighted scores, for Alternative 1, its weighted score is: $0.12*0.14 + 0.52*0.35 + 0.06*0.19 + 0.30*0.45 = 0.3452$. By the same way, we can get the weighted scores of Alternative 2 and Alternative 3, Finally collecting them all in Table 8.

At this point, through Bayesian Neural Network classification with Comparative Scoring Quantification Method AHP, it concludes that investment Alternative 3 is what

Table 7. Weights of Benefit 1–3

Benefit 2	Alternative 1	Alternative 2	Alternative 3
reduced mortality risks	7%	5%	8%
W_{B2}	**0.35**	**0.25**	**0.4**
Benefit 3	Alternative 1	Alternative 2	Alternative 3
increased work productivity	300,00	750,000	550,000
W_{B3}	**0.19**	**0.47**	**0.34**
Benefit 4	Alternative 1	Alternative 2	Alternative 3
Health care cost savings	250,0000	100,0000	200,0000
W_{B4}	**0.45**	**0.18**	**0.37**

Table 8. Overall Weighted Score Matrix

Z	W	Alternative 1	Alternative 2	Alternative 3
Benefit 1	0.12	0.14	0.57	0.28
Benefit 2	0.52	0.35	0.25	0.40
Benefit 3	0.06	0.19	0.47	0.34
Benefit 4	0.30	0.45	0.18	0.37
Weighted Score		**0.3452**	**0.2806**	**0.373**

we should implement. The advantage of this AHP Multi-Attribute Decision Making (MADM) is that we do not abandon consideration of any 'Benefit', at the same time also not against technical and moral difficulties to forcibly monetize benefit 1 and benefit 2. More importantly, it can be justified because we assigned a 52% weight to benefit 2 'reduced mortality risks', which means saving human lives is more important than all other benefits combined, which speaks to the morality of the decision. And by giving a 30% weighting to Benefit 4 'health care cost savings', we ensure that the investment objective is primarily considered in decision-making.

2.3 Limitations and Improvement

The Multi-Attribute Decision Making (MADM) with AHP method can solve the ethical issue associated with monetization in Benefit valuation of BCA. However, the above application case has a major flaw, which is that it does not take into account the "cost". Recalling the problem, I mentioned at the beginning of this paper, the investment has two kinds of costs. Cost 1 is the sum of costs that can be monetized. It can be a combination of monetary costs, human and equipment costs, and opportunity costs. Cost 2 is the cost that cannot be monetized, such as the normative issues brought to society, and the negative impact on religion and culture. One example is the Manhattan Project, which can consider all the human, material, time input into the cost of building an atomic bomb. But there is another cost, which is the dehumanization of such weapons of mass destruction. 'Dehumanization' is an essential cost that cannot be rightly measured by either moral or technical means. (In philosophy, we have a tradition of valuing humanity above individual lives). The implication is that Cost 1 and Cost 2 are two completely different types of attributes. For 'Cost 1' we can simply modify the AHP-MADM above to take it into account. Concretely, we just set Cost 1 as an attribute of the same type as the other four benefits attributes, but in the opposite direction that means the smaller the original value of Cost 1, the higher the score of it in the matrix. Or realizing the positive transformation of 'cost' type of attributes via formula (3).

$$\text{Attribute}'_{\text{Cost1 i}} = \frac{\text{Attribute}_{\text{cost1 max}}}{\text{Attribute}_{\text{cost1i}}} i = 1, 2, 3, \ldots \tag{3}$$

However, for Cost 2, we cannot simply add it to the judgment matrix as an attribute of the same kind and opposite direction. Because we have already argued that it is a completely different kind of attribute. In this example, suppose 'Cost 2' is the negative effect of unfair distribution of social benefits or the worsening of the gap between the rich and the poor. We can think that the essence of Benefit1- 4 and Cost 1 is measurement of efficiency, while the essence of Cost 2 is measurement of equity. AHP with efficiency as the decision criterion cannot include the attribute of equity at any layer, and vice versa. Therefore, to take both efficiency and equity into account in decision making, we need to change the above AHP model to the new one as Fig. 3.

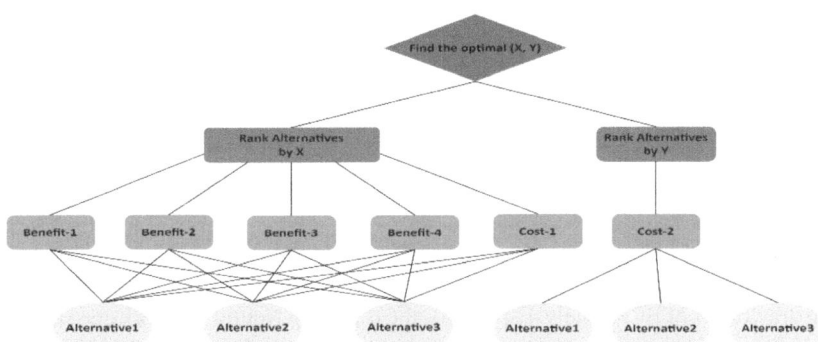

Fig. 3. New AHP-MADM Paradigm

The above model implies that the final output layer is a two elements array [X, Y], as there are two perceptron. This means that we must assign an additional coefficients (weight) to the two output values X,Y and then integrate them into a single value in order to make a decision ranking. Here instead of adopting the most common method Expected Utility Theory (EUT) [9], I prefer the Kahneman and Tversky (1979) "Prospect Theory"[10]. The theory suggests that 1) because of culture, religion and education, people intuitively hold psychological preferences when making decisions. One such Preference that has been proved by psychological experiments is "Fairness Preference". It refers to the decision maker's psychological preference for the benefits distribution, which will influence the decision maker's refusal behavior on the unfair result [3]. 2) under certain risk conditions, people may have different preferences for risk (negative) and return (positive) attributes. For example, people may care more about the reduction of losses or be more sensitive to changes in losses, even if the potential gains could compensate for the loss. 3) each person's psychological reference point is different, so the utility measure of "gain" and "loss" is different [10]. The Prospect Theory suggests that a preference coefficient k should be added to the expected utility formula. And the decision criterion is no longer the merely maximization of aggregate utility, but the maximization of aggregate utility while avoiding negative effects as much as possible. Topologically, it is to find a decision point that is closer to the positive ideal point than other points, and further away from the negative ideal point as illustrated by Fig. 4. This can solve the issue of multiple solution.

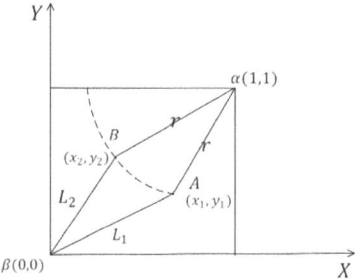

Fig. 4. Optima Solutions Algorithm in Geometric Expression

3 Conclusion

Through a real-world health business decision-making scenario analysis, the study has shown the combination of MADM with BCA decision-making framework can overcome several drawbacks of BCA solely. Especially, the Bayesian Neural Network classification is employed to implement a Comparative Scoring Quantification Method that integrates prior judgment and posterior data, thereby addressing the challenge of monetizing intangible value. Furthermore, the article also proposes enhancement to the aforementioned MADM by introducing Prospect Theory, which allows each enterprise to have its own unique decision preferences. This can be reflected in the setting of weight parameters in the finial utility function. Although MADM has many advantages as a comprehensive decision-making method, its operation is often complex. And the validity of this method needs to be verified by huge amounts of real cases. Nevertheless, the two challenges of complexity and validity are being overcome by the advancements in data engineering and machine learning today. By leveraging big data from business history, which encompasses a wealth of successful and failed decision-making cases from the past, we can calibrate our parameters and train our MADM model to be more precise and stable which can sever an important role in Business Intelligence.

Acknowledgments. I appreciate the comprehensive guidance about BCA from Kerry Krutilla, Professor of Indiana University. The business case scenario analysis in the paper is deeply inspired by him. As well as appreciate Frank Page Professor of LSE. His advice on Bayesian Networks Algorithm Application in Finance field was very valuable.

Disclosure of Interests. The authors declare that there are no conflicts of interest in this work.

References

1. Smith, C., Neumann, J., Morgenstern, O.: Theory of Games and Economic Behavior. Journal of the American Statistical Association **40**, 263 (1945)
2. Kahneman, D., Tversky, A.: Decision, probability, and utility: Prospect theory: An analysis of decision under risk. (1979)
3. Opricovic, S.: Multicriteria optimization of civil engineering systems. PhD Thesis, Faculty of Civil Engineering, Belgrad (1998)
4. Stewart, T.J.: A Critical Survey on the Status of Multiple Criteria Decision Making Theory and Practice. Omega-international Journal of Management Science **20**, 569–586 (2022)
5. Louviere, J.J., Hensher, D.A., Swait, J.D.: Stated Choice Methods: Analysis and Applications. (2000)
6. Williams, C.K., Barber, D.: Bayesian Classification With Gaussian Processes. IEEE Trans. Pattern Anal. Mach. Intell. **20**, 1342–1351 (1998)
7. Kwon, Y., Won, J., Kim, B., Paik, M.C.: Uncertainty quantification using Bayesian neural networks in classification: Application to biomedical image segmentation. Comput. Stat. Data Anal. 142 (2020)
8. Ye, J.: A multicriteria decision-making method using aggregation operators for simplified neutrosophic sets. J. Intell. Fuzzy Syst. **26**, 2459–2466 (2014)
9. Smith, C., Neumann, J., Morgenstern, O.: Theory of Games and Economic Behavior. J. Am. Stat. Assoc. **40**, 263 (1945)
10. Forsythe, R., Horowitz, J.L., Savin, N.E., Sefton, M.: Fairness in Simple Bargaining Experiments. Games Econom. Behav. **6**, 347–369 (1994)

Exploring Vulnerabilities in Near Field Communication (NFC) Devices: A Comprehensive Investigation

Sphamandla Sangweni and Khutso Lebea[✉] [ID]

University of Johannesburg, Johannesburg, South Africa
223166580@student.uj.ac.za, klebea@uj.ac.za

Abstract. Near Field Communication (NFC) is gaining increasing attention as a highly convenient, short-range, contactless communication method. It has garnered interest primarily due to its user-friendly support for mobile payment, and it has been widely used in various mobile devices for some time now. The popularity of NFC is on the rise, thanks to the rapid proliferation of NFC-enabled devices in the market. This technology facilitates data transfer between devices when brought into proximity. It is primarily designed for seamless integration with mobile phones, enabling them to communicate with other phones through peer-to-peer connections and read or write information on NFC tags and cards using a compatible reader. Apple Pay, Google's Wallet, Samsung Pay, and Cash App have successfully integrated NFC technology for mobile transactions. However, as with any wireless technology that experiences widespread adoption, NFC is not immune to security concerns. It can be vulnerable to several security-related attacks, such as a man-in-the-middle, denial of service (DOS), and relay attacks. This paper explores the vulnerabilities associated with NFC and the different attacks to which NFC is susceptible.

Keywords: Near Field Communication · Mobile Payment

1 Introduction

Near Field Communication (NFC) has rapidly evolved into a cutting-edge wireless communication technology that operates at a frequency of 13.56 MHz, a development that originally stemmed from the foundational principles of Radio Frequency Identification (RFID). Unlike many other wireless communication technologies, NFC is characterised by its short-range communication capability, typically operating within a range of a few centimetres. This short-range interaction prioritises low bandwidth while ensuring high frequency, particularly suited for specific applications that demand proximity and low power consumption (Coşkun et al., 2011). The unique properties of NFC, combined with its versatility, have contributed to its widespread adoption in various industries and consumer applications, from mobile payments to public transportation and beyond.

As NFC technology becomes increasingly integrated into everyday devices, such as smartphones and contactless payment systems, security becomes more prominent. The

K.-W. Huang et al. (Eds.): ICFT 2024, CCIS 2437, pp. 56–66, 2025.
https://doi.org/10.1007/978-981-96-3811-6_6

convenience that NFC offers—such as swift and seamless transactions—can sometimes overshadow the necessary consideration of security aspects, leaving potential vulner-abilities unaddressed. This is especially critical as industries with sensitive data, like banking and eHealthcare, are embracing NFC for contactless payments, identity verifi-cation, and data transmission tasks. While NFC's ease of use is a significant advantage, adopting this technology without adequate security safeguards can lead to serious risks, given its wireless nature and the sensitive information often involved in transactions.

A prime example of NFC's growing influence is its application in mobile phones, where it is widely used to facilitate wireless payments through platforms such as Google Pay and Apple Pay. However, with the increased use of NFC technology, the potential for security threats grows in tandem. This leads to the central problem addressed in this paper: NFC, a wireless communication network, is susceptible to various attacks, much like other wireless technologies. One of the most concerning threats is the relay attack, in which an attacker intercepts and forwards NFC communications between two parties without their knowledge. In such attacks, an unauthorised third party can manipulate or gain access to the transferred data, posing serious user risks, particularly in financial transactions.

In addition to relay attacks, another significant security risk is eavesdropping. This occurs when an attacker covertly listens to the communication between two wireless devices, such as during an NFC transaction, to steal sensitive information (Riyazuddin, 2011). These attack types underscore the vulnerabilities inherent in NFC technology, especially when used in financial applications. NFC's convenience to consumers and businesses must be carefully balanced against the potential risks posed by these and other forms of attacks.

To further explore the real-world implications of NFC vulnerabilities, this paper will examine these security risks through a comprehensive case study. The case study will focus on a financial application scenario where NFC-based systems are at risk of attack, shedding light on how these threats manifest in practical settings. Through this exploration, the paper aims to clarify the risks and emphasise the need for more robust security measures as NFC technology increases across industries. By analysing the technology's benefits and associated risks, this research contributes to a broader conversation about how to securely implement NFC systems in an increasingly digital world.

Section 2 outlines the case study. Section 3 explores the background of NFC technol-ogy, providing a comprehensive foundation for the paper and drawing connections to the case study. Section 4 looks at the security threats NFC-based systems face, and Sect. 5 explores some solutions. The paper is concluded in Sect. 6, where potential research improvements and highlights of the paper are presented.

2 Case Study

A scenario involving a user and their NFC-enabled device is introduced to better under-stand the security vulnerabilities of NFC wireless connectivity. Imagine a typical shop-per who uses their smartphone to pay at their local grocery store. This Android device boasts significant computational capabilities, supporting contactless payments through

NFC technology. It is equipped with a Secure Element (SE), as described by Fang et al. in 2017, in addition to its operating system, processor, and memory.

However, a potential threat exists during the payment process: a malicious attacker could execute a relay attack. In this attack, the attacker may install a malicious application on the user's phone that allows the attacker to steal the user's information passed through the network.

2.1 Scenario

John is a regular shopper who has just completed his purchase at the nearby grocery store. With his NFC-capable Android smartphone, John makes his payment by tapping it on the point of sale (POS) machine, utilising Android Pay to facilitate the transaction.

However, it is important to note that John might not know about a potential security risk on his phone. An application running in the background could listen for specific commands, known as the Application Protocol Data Unit (APDU), over a network connection. These commands are then passed on to a secure element for processing, and the responses are sent back using the same network connection.

On the other end of this network connection, a card emulator forwards these APDU commands and responses to and from a smart card reader. This smartcard reader could be a component of various systems, such as an access control system or a POS terminal.

This situation highlights the urgent need to identify and address security vulnerabilities related to NFC wireless technology.

3 Background

This section focuses on the history and background of NFC and how the technology has evolved from its origins, RFID. Then, the different operating modes and the use cases of NFC are detailed concerning the case study in Sect. 2.

While NFC technology includes basic security measures built into its protocol, they are relatively minimal and leave room for enhancement to create a more robust and secure NFC architecture. According to Coşkun et al. (2011), "NFC is based on the ISO/IEC 14443 standard and operates in the 13.56 MHz Radio Frequency (RF) spectrum." This standard, widely accepted across industries, ensures NFC devices and systems interoperability. Additionally, NFC supports data transfer rates of 106 Kbps, 212 Kbps, and 424 Kbps, making it efficient for short message transfers. These three transmission rates allow NFC-enabled devices to quickly exchange data within seconds, a feature that suits the technology, particularly for cardless transactions, such as mobile payments and contactless ticketing. The "tap and go" nature of NFC transactions exemplifies this speed and convenience, which are essential in high-traffic environments like retail or public transportation.

Despite its advantages, NFC technology often involves transmitting highly sensitive and confidential information, including payment credentials, transaction details, security protocols, authentication and authorisation data, and device-specific information. This raises the critical concern of whether the current security measures can protect such sensitive data as it traverses the airwaves. Although the protocol implements encryption and other protective measures during data exchanges, the travel medium (wireless communication) inherently poses risks, particularly in environments where malicious actors might attempt to intercept or manipulate data.

One of the primary reasons NFC was designed with a lightweight architecture was to ensure its seamless integration with devices with limited hardware resources, such as smartphones, wearables, and other NFC-enabled gadgets (Coşkun et al., 2011). This focus on simplicity allows NFC to be easily adopted across various devices. Still, it also means that the built-in security features are not as comprehensive as they might be in more resource-heavy systems. While encrypted data packets offer a degree of protection, NFC communications are still susceptible to various security attacks, including eavesdropping (where an attacker listens in on the communication), sniffing (capturing data being transmitted), data manipulation, man-in-the-middle attacks, and relay attacks. These vulnerabilities highlight the need for more advanced security measures in NFC-based systems, especially when regularly exchanging sensitive information.

As NFC continues to grow in popularity, particularly in industries that handle sensitive transactions like finance and healthcare, improving its security features has become a pressing concern. The lightweight design that facilitates ease of use and quick transactions must be balanced with enhanced security protocols to prevent unauthorised access and exploitation of the technology. This balance is critical to ensuring that NFC can be widely adopted without compromising the privacy and security of the users.

3.1   History of Near Field Communication

NFC has gained prominence due to its unique characteristics, including short-range communication, low bandwidth, and high frequency (Coşkun et al., 2011).

Radio Frequency Identification (RFID) technology is like NFC. However, it has a more extended transmission range than NFC and has been utilised in mobile payments in Japan and South Korea, two prominent countries. However, due to the extended transmission range, certain RFID-enabled mobile payments are deemed less secure when compared to NFC-enabled payments, as highlighted by Hayashi (2012) and Coşkun et al. (2011). Two main differences between RFID and NFC are scanning distance and the method of communication (Singh et al. 2018).

According to Coşkun et al. (2011), NFC technology has applications beyond mobile payments. As more mobile phones become NFC-enabled and commercial services are introduced, individuals will be able to use their phones to pay for goods and services, access hotel rooms or apartments, update their social network information, upload their health data to hospital monitoring systems from their homes, and enjoy a host of other services. In the case study presented in Sect. 2, it is observed that one of the applications where a user uses their mobile phone to pay for goods at the point-of-sale machine.

The NFC system operates in two distinct modes, like RFID technology: active and passive. Both NFC devices independently generate radio frequency signals to transmit data in active mode. Conversely, in passive mode, only one NFC device initiates the radio frequency field. Coşkun et al. (2011) study describes that NFC can facilitate two-way or peer-to-peer communication and a reader/writer mode. The case study represents the latter mode, where one device (the mobile phone) emulates the behaviour of a contactless smart card, and the POS terminal acts as the reader/writer.

3.2 NFC Operation Modes

NFC's three primary operation modes are Card Emulation Mode, Reader/Writer Mode, and Peer-To-Peer Mode. This paper explores how NFC-enabled devices function in each mode and the various applications they serve, from mimicking smart cards to facilitating data transfer. Additionally, this section will discuss the two communication modes in NFC, two-way and one-way, and how they shape the interaction between NFC devices. This comprehensive overview will provide a clearer understanding of NFC's versatility and the diverse scenarios in which it plays a pivotal role in modern technology and daily life.

NFC technology offers three distinct operation modes:

- **Card Emulation Mode:** In this mode, an NFC-enabled device can replicate the functionality of smart cards. Users can employ this mode for various transactions, including purchases, ticketing, and access control for public transportation (Singh et al., 2018).
- **Reader/Writer Mode:** NFC-enabled devices can read information from NFC tags embedded in smart posters and various objects. This functionality can result in ticket cloning, where individuals duplicate tickets to gain discounts on purchases. If the cloned ticket passes the verification process, it remains valid until expiration. The method of ticket cloning can vary depending on the design of the ticketing system, but the core goal is to exploit the ticket's benefits for as long as possible (Singh et al., 2018). In Reader/Writer mode, modern NFC-enabled devices can read NFC tags such as contactless smart cards and RFID tags. When a tag is nearby, it is automatically detected, allowing the device to either read data from or write data to the tag. One significant application of this mode is smart posters, where users can interact directly with promotional or informational content (Singh et al., 2018).
- **Peer-To-Peer Mode:** Two powered devices can utilise a peer-to-peer mode specific to NFC, which enables direct communication between the devices as if they were part of a network. This mode facilitates data exchange between the devices without additional infrastructure or intermediaries.
- **Card Emulation Mode (Revisited):** An NFC device functions as an NFC card, operating passively. In this mode, the smartphone does not generate its own RF field; instead, the NFC reader produces it. An external NFC reader, such as a point-of-sale (POS) terminal, can then interact with and access the emulated NFC card.

3.3 Communication Modes

Communication in NFC can occur in either active or passive modes, depending on the device's capabilities and power source:

- **Two-Way Communication**: Devices equipped with NFC can both read from and write to each other, allowing for seamless data exchange. For example, two Android devices can be connected using NFC to transfer data such as contacts, links, or photos (Shobha et al. 2016).
- **One-Way Communication:** A powered device, such as a smartphone, credit card reader, or commuter card terminal, reads from and writes to an NFC chip. For instance,

when a commuter card is tapped on a terminal, the NFC-powered terminal deducts money from the card's balance (Shobha et al. 2016). NFC Security Challenges

The convenience of NFC technology is well illustrated by platforms like Android Pay, which allow users to make quick, contactless payments with a simple tap. However, as technology evolves, so does the sophistication of potential security threats. One of the most pressing concerns in this context is the rise of relay attacks, a security breach where attackers intercept and manipulate the communication between two NFC-enabled devices. This section delves into the specifics of relay attacks, examining their mechanics, potential consequences, and the tools attackers need to execute them. Understanding these elements is crucial for addressing the growing challenges in NFC security and developing more robust defences against these exploits.

With the widespread adoption of NFC technology, particularly in mobile payment systems such as Android Pay, vulnerabilities have come to light, offering malicious actors opportunities to exploit them. Among these vulnerabilities, relay attacks have emerged as a particularly dangerous threat. As Riyazuddin (2011) explains, relay attacks occur when an attacker intercepts and relays NFC communication between two devices without either party's knowledge. By doing so, attackers can manipulate the data exchange and potentially initiate unauthorised transactions. This is especially concerning in scenarios involving financial data, where the implications of such attacks can be devastating, leading to financial losses and compromised personal information. The case study outlined in this paper highlights a practical example of such an attack, demonstrating how users like John can become victims of unauthorised transactions through no fault of their own.

Relay attacks are a well-known vulnerability in NFC technology and in wireless payment systems in general, particularly when it comes to NFC and Secure Element technology (Francis et al., 2010). To effectively execute a relay attack, the attacker requires three essential components:

1. The reader device, or "mole", is positioned close to the victim's NFC-enabled card or smartphone. It captures the communication signals intended for the legitimate reader, initiating the first step of the attack.
2. The card emulator device, or "proxy": The proxy device serves as an intermediary between the legitimate card reader and the attacker. It emulates the behaviour of the legitimate NFC card or device, fooling the legitimate reader into thinking it interacts with the actual user.
3. An efficient communication channel: The attacker must establish a fast, reliable connection between the mole and the proxy. This allows for real-time relaying of data, ensuring that the legitimate card reader continues to communicate without any noticeable delay or disruption, making the attack difficult to detect in real-time.

These components work together to create a seamless, undetectable relay of information between the victim's NFC device and the legitimate reader. While the attack occurs, the victim is often unaware that their device's communication has been compromised, and the legitimate reader continues processing the transaction as if nothing is amiss. The potential consequences of such attacks are significant, especially in environments where NFC is used for sensitive transactions, such as banking, retail, and public transportation.

Understanding the mechanics of relay attacks is essential for developing stronger defences against them. As NFC technology becomes more integral to modern commerce and communication, the risk of these attacks will likely increase, necessitating more sophisticated security measures to safeguard user data and financial information. Efforts to mitigate the threat of relay attacks include improving encryption protocols, implementing real-time transaction monitoring, and utilising multi-factor authentication to ensure that only authorised users can complete transactions. These measures are critical to fortifying NFC systems against the evolving landscape of digital threats.

4   Security Threats in NFC-Based Systems

Several attacks target NFC-based systems, with eavesdropping being a common one. Eavesdropping happens when an attacker intercepts communication between two NFC-enabled devices. Although NFC operates over short radio waves, attackers with the right equipment can capture these signals and access sensitive information. The short range of NFC makes it harder for attackers, but it's still possible if they can get close to the devices. Encrypting NFC communications prevents eavesdropping (Alzahrani et al., 2013).

Another frequent threat is the relay attack, where an attacker extends NFC's communication range to make unauthorised transactions. The attacker uses two devices: one near the victim's NFC-enabled device and another near the payment terminal. By relaying communication between these two points, the attacker can deceive the terminal into thinking the transaction is legitimate. This type of attack shows the need for extra security measures beyond proximity, such as biometric verification and transaction confirmation (Tu & Piramuthu, 2020).

Data modification is another serious threat in NFC-based systems. In this attack, an attacker intercepts and changes the data transmitted between devices. For instance, the attacker might alter the transaction amount during a mobile payment transaction, leading to fraudulent charges. To guard against data modification, implementing strong encryption and data integrity checks is essential. These security measures help detect and reject tampered data, immediately flag any changes and prevent financial losses and fraud (Chattha, 2014).

NFC tag cloning is a common attack in systems that use NFC for access control or payments. Attackers copy the data from a genuine NFC tag to create a fake one, which can then be used to gain unauthorised access or make fraudulent transactions. To combat this risk, NFC systems should use cryptographic signatures and unique identifiers for each tag. This approach ensures that even if the data is copied, the cloned tag cannot be used without the correct cryptographic key, thus enhancing security (Leclerc & Kärrström, 2022).

4.1   Relay Attack in Theory

This section will investigate relay attacks and their significance in security. Building upon the analogy of a chess novice playing against two masters, the section explores how relay attacks work in practice and the potential vulnerabilities they exploit in security protocols. By understanding the mechanics and implications of relay attacks, one can better grasp the importance of safeguarding against them in modern cybersecurity practices.

In a study by Francis et al. in 2010, they described the theory of a relay attack as a grandmaster chess problem game. When someone who does not know how to play chess goes up against two chess masters, this person starts two chess games, one with each master, and tells each master what moves the other is making. It turns out that even though both masters think they are playing against a beginner, they are competing against each other.

In a security context, this scenario parallels what is known as a relay attack. A malicious actor can circumvent security protocols through this attack by merely relaying challenges and responses between two legitimate entities. The security protocol is successfully executed as the attacker consistently can provide the correct response, acquired by forwarding the original message and recording the subsequent response. Consequently, both participating parties consider the attacker to be a legitimate participant in the protocol.

Crucially, in this scenario, the attacker is never required to know the specifics of the information being relayed. This means the attacker does not need to understand the protocol's structure, the algorithms employed, the content of plaintext data exchanged, or any confidential key material (Francis et al., 2010).

Implementation

This sub-section explores the practical implementation details of relay attacks, drawing from the research of Chen et al. (2014) and Roland et al. (2012). Specifically, an investigation into how these attacks can be executed using Application Protocol Data Unit (APDU) commands, drawing to our scenario of a shopper, John, at a grocery store. The section will focus on a proposed relay attack scenario, where an application installed on the victim's mobile phone is the relay point for APDU commands. It will examine the mechanics of this attack, including the transmission of commands to a secure element and the use of a card emulator on the other end to interact with a smart card reader, potentially embedded in access control systems or point-of-sale terminals.

Understanding these implementation specifics is crucial for comprehending the practical implications and countermeasures against relay attacks in modern security contexts.

Chen et al. (2014) have discussed in their research how a relay attack can be accomplished using Application Protocol Data Unit (APDU) commands. Roland et al. (2012) also emphasise the utilisation of APDU commands to carry out such attacks.

In our proposed relay attack scenario, we envision a relay application running on the victim's (John's) mobile phone. This application actively awaits APDU commands via a network socket and subsequently relays these APDUs to a secure element. The resulting responses are then transmitted back through the same network socket. A card emulator takes charge on the opposite end of this network connection, forwarding the APDU commands (along with the responses) to and from a smart card reader. This smartcard reader might be integrated into various systems, such as an access control mechanism or a point-of-sale terminal (as described by Roland et al. in 2012).

Unlike traditional relay scenarios operating at the data link layer, our attack scenario relays command and response packets (APDUs) at the application layer. This means we are dealing with a higher-level protocol, and there are hardly any strict timing requirements for the relay channel. This flexibility allows us to use various communication channels like Bluetooth, Wi-Fi, or the mobile phone network as relay channels (Roland et al. 2012).

5 Existing Security Solutions for NFC Technology

Several existing security solutions address vulnerabilities in NFC-based systems. Encryption is one of the most common methods used to protect NFC communications. It works by encoding the data before transmission, ensuring that even if an attacker intercepts the communication, they cannot read or alter the information without the correct decryption key. End-to-end encryption is especially effective for securing sensitive transactions, such as mobile payments and e-health data exchanges, from unauthorised access (Govender & van Niekerk, 2021).

Multi-factor authentication (MFA) is another crucial security measure for NFC-based systems. MFA requires users to confirm their identity through multiple methods, such as a password, a fingerprint, and an NFC-enabled device. This approach makes it much harder for unauthorised users to gain access. Even if attackers intercept NFC communication or clone a device, they need additional authentication factors to complete the transaction. By using several layers of verification, MFA provides a higher level of security than relying on just one method (Singh, Adzman, & Hassan, 2018).

Tokenisation is a powerful tool for securing NFC-based payments. In this process, sensitive data, like credit card numbers, is replaced with a unique token with no real value. This token is used during transactions instead of the actual payment information. Even if attackers intercept the token, they cannot use it for fraudulent transactions. When combined with encryption and multi-factor authentication (MFA), tokenisation provides a robust security solution for mobile payments and other NFC-based services (Ozdenizci, Ok, & Coskun, 2016).

Blockchain technology offers a promising solution for securing NFC communications, especially in financial systems. It works by storing patient data on a decentralised, tamper-resistant ledger, making it difficult to alter the information without proper authorisation. This method adds an extra layer of security to NFC-based financial applications, protecting sensitive customer data from unauthorised access and modification. Although blockchain adoption is still emerging, it has significant potential to enhance NFC security in critical areas like in fintech (Alzahrani & Bulusu, 2018).

6 Conclusion

Near Field Communication (NFC) has undeniably revolutionised the way we interact with technology and the convenience it brings to our daily lives. Its seamless integration into mobile payment systems, access control, and data transfer between devices has made it ubiquitous today. However, as with any technology, NFC is not without its vulnerabilities, and this comprehensive investigation has shed light on the security challenges it faces.

This paper has delved deep into the world of NFC, exploring its origins in Radio Frequency Identification (RFID) and its evolution into a powerful and versatile communication technology. We have examined its various operation modes, including Card Emulation Mode, Reader/Writer Mode, and Peer-To-Peer Mode, highlighting the vast array of applications NFC serves. Moreover, we have explored the two communication modes, two-way and one-way, that define how NFC devices interact with each other.

One of the most significant concerns discussed in this paper is the vulnerability of NFC to relay attacks. We have dissected the mechanics of relay attacks, emphasising their potential consequences and the crucial components required for their execution. By understanding how relay attacks work, we can better appreciate the pressing need for robust security measures in NFC technology.

In our case study, we illustrated a scenario where a user's NFC-enabled device could be compromised during a routine transaction, emphasising the real-world implications of these vulnerabilities. This example is a stark reminder of the importance of promptly addressing NFC security concerns.

As we look to the future, it becomes evident that NFC technology will continue to play an essential role in our lives. Its potential applications are vast and ever-expanding, from simplifying payments to enhancing access control and data sharing. However, to fully embrace NFC's advantages, we must address its security challenges head-on.

This paper has provided a comprehensive overview of NFC technology, its operation modes, communication modes, and the security challenges, focusing on relay attacks. It is clear that NFC's potential is vast, but its security must be robust to protect users and ensure its continued success. As we move forward, researchers and industry professionals must work collaboratively to develop and implement effective security solutions, safeguarding the promise of NFC in our increasingly interconnected world. NFC's potential is enormous, and by addressing its vulnerabilities, we can unlock even more of its capabilities while ensuring a secure and trustworthy user experience.

References

Alzahrani, N., Bulusu, N.: Block-supply chain: a new anti-counterfeiting supply chain using NFC and blockchain. In: Proceedings of the 1st Workshop on Cryptocurrencies and Blockchains for Distributed Systems, pp. 30–35 (2018)

Alzahrani, A., Alqhtani, A., Elmiligi, H., Gebali, F., Yasein, M.S.: NFC security analysis and vulnerabilities in healthcare applications. In: 2013 IEEE Pacific Rim Conference on Communications, Computers and Signal Processing (PACRIM), pp. 302–305. IEEE (2013)

Chattha, N.A.: NFC – vulnerabilities and defense. In: 2014 Conference on Information Assurance and Cyber Security (CIACS), pp. 35–38. IEEE (2014)

Chen, C.H., Lin, I.C., Yang, C.C.: NFC attacks analysis and survey. In: The Eighth International Conference on Innovative Mobile and Internet Services in Ubiquitous Computing, pp. 458–462. IEEE (2014)

Coşkun, V., Ok, K., Özdenizci Köse, B.: Near Field Communication: From Theory to Practice (2011)

Fang, H., Liu, X., Yang, B.: A countermeasure against relay attack in NFC payment. In: Proceedings of the Second International Conference on Internet of Things, Data and Cloud Computing (2017). https://doi.org/10.1145/3018896.3025144

Francis, L., Hancke, G., Mayes, K., Markantonakis, K.: Practical NFC peer-to-peer relay attack using mobile phones. In: Radio Frequency Identification: Security and Privacy Issues: 6th International (2010)

Hayashi, F.: Mobile payments: what's in it for consumers? In: Economic Review-Federal Reserve Bank of Kansas City, p. 35 (2012)

Govender, C., van Niekerk, B.: Secure key exchange by NFC for instant messaging. In: 2021 Conference on Information Communications Technology and Society (ICTAS), pp. 27–33. IEEE (2021)

Leclerc, S., Kärrström, P.: Cloning Attacks Against NFC-Based Access Control Systems (2022)

Ozdenizci, B., Ok, K., Coskun, V.: A tokenisation-based communication architecture for HCE-enabled NFC services. Mob. Inf. Syst. **2016**(1), 5046284 (2016)

Riyazuddin, M.: NFC: a review of the technology, applications, and security. ABI Research (2011)

Roland, M., Langer, J., Scharinger, J.: Practical attack scenarios on secure element-enabled mobile devices. In: 2012 4th International Workshop on Near Field Communication, pp. 19–24. IEEE (2012)

Shobha, N.S.S., Aruna, K.S.P., Bhagyashree, M.D.P., Sarita, K.S.J.: NFC and NFC payments: a review. In: 2016 International Conference on ICT in Business Industry and Government (ICTBIG), pp. 1–7. IEEE (2016)

Singh, M., Mandy, M., Adzman, K., Hassan, R.: Near field communication (NFC) technology security vulnerabilities and countermeasures. Int. J. Eng. Technol. **7**, 298–305 (2018). https://doi.org/10.14419/ijet.v7i4.31.23384

Tu, Y.J., Piramuthu, S.: On addressing RFID/NFC-based relay attacks: an overview. Decis. Support. Syst. **129**, 113194 (2020)

A Novel Electronic Payment System Based on Zero-Knowledge Proof and Blockchain

Viet-Thang Nghiem, Thi-Huong Tran, and Ba-Lam Do[✉]

Hanoi University of Science and Technology, Hanoi, Vietnam
thang.nv190088@sis.hust.edu.vn, {huong.tranthi,lam.doba}@hust.edu.vn

Abstract. As the global economy continues to experience rapid digital transformation, electronic payment (e-payment) has emerged as a pivotal driver of financial transactions worldwide. In this context, ensuring the security and privacy of electronic payment transactions has become a paramount concern. In this paper, we introduce a novel method of e-payment that incorporates a zero-knowledge proof scheme and blockchain technology to enhance privacy and bolster trust between parties engaged in financial transactions. In particular, we rely on advanced cryptographic techniques to verify transaction information without exposing sensitive payment details such as customers' credit card numbers. We implemented the proposed system to evaluate the performance and effectiveness. The results show that our approach is promising.

Keywords: Online payment · Zero-Knowledge proof · Blockchain

1 Introduction

Electronic payment, often abbreviated as e-payment, is a modern and convenient method of conducting financial transactions through digital channels. It allows individuals and businesses to transfer money electronically, eliminating the need for physical cash or paper-based transactions. With the widespread adoption of digital technology and the internet, e-payment has become an integral part of the global economy. Especially during the COVID-19 pandemic, people's awareness and the use of electronic commerce have increased. In this context, we need a reliable, trusted payment system. Such a system must allow the transacting parties to confirm each other's identity, ensuring that no third parties have access to that sensitive information [3,19].

Several e-payment methods exist, such as traditional card-based payments (credit cards and debit cards), digital wallet payments, mobile money payments, and online banking payments. In particular, card-based payment is one of the most used e-payment methods. Yet, it is invulnerable to attack from malicious third parties such as phishing or stealing, resulting in financial losses to both customers and financial organizations [14,16]. Hence, developing a new method that could make customers' card information more private and secure is very

K.-W. Huang et al. (Eds.): ICFT 2024, CCIS 2437, pp. 67–75, 2025.
https://doi.org/10.1007/978-981-96-3811-6_7

crucial for a safer commercial environment. Such a method must have the ability to allow users to hide their card identity while still letting other merchants verify their trustworthiness.

In this paper, we introduce a novel e-payment system that combines Zero-knowledge proof and blockchain to enhance the security and privacy of card payment transactions. Zero-knowledge proof [8,10] is a modern cryptography that allows one party (i.e., the prover) to prove the validity of a statement to another party (i.e., the verifier) without revealing any specific information about the statement itself. This ingenious technique ensures that the prover (i.e., customer) can prove their knowledge of specific data (i.e., cards' information) to the verifier (i.e., the card issuer or merchant if needed). Following this idea, Zero-knowledge succinct non-interactive argument of knowledge (ZK-SNARK [11,17]) is an algorithm that enables the prover to demonstrate the validity of a statement to the verifier without engaging in any back-and-forth communication. ZK-SNARKs eliminate the need for interactive communication between the prover and the verifier, significantly reducing computation and transmission overhead while allowing multiple verifiers to verify the same proof. The ability to produce compact proofs makes ZK-SNARKs highly efficient and suitable for resource-constrained environments like mobile devices and blockchain networks. In addition, we make use of blockchain [15,18] as a way to identify customers and archive their corresponding receipts information. Each customer will be assigned a unique wallet. Furthermore, each time a payment is completed, customers will perform a transaction that stores the receipt information on the network. This information is immutable, and such transactions can only be sent from the corresponding customer's address.

The remainder of this paper is organized as follows. In Sect. 2, we introduce the fundamental knowledge that our research relies on. Section 3 presents the design and implementation of our system. Section 3.3 illustrates the system using detailed experiments. Finally, we conclude with an outlook on future research in Sect. 5.

2   Background Knowledge

2.1   Payment Process

Each online payment is conducted via two main stages, including (i) authorization and (ii) clearing and settlement [9]. The first stage aims to verify customer card information, such as a valid card and sufficient balance, to ensure that the transaction can be completed. Whereas the second stage allows the debit to the customer account and then credit to the merchant account. Usually, the merchant suggests that their customers store the cards directly on the merchant's application. On the one hand, customers can save time when submitting their card details. On the other hand, this is the case that customer cards could be vulnerable because of attacks and information leaks.

To solve this problem, EMV (Europay, Mastercard & Visa) created EMV® Payment Tokenization [7], which enhances transaction security by replacing the most valuable data to a fraudster within that transaction - the primary account number (PAN) - with an alternative value: a payment token. This method uses cryptographic technology to ensure that other malicious parties cannot reconstruct the PAN from the token itself. EMV® Payment Tokenization offers both cardholders and merchants a significant gain of security and protection of payment information throughout a transaction, reducing the risk of financial loss. Along with these advantages, EMV® Payment Tokenization also has some downsides. Both merchant and customer may be confused while using payment tokens for transactions. The reason is that the PAN has been entirely replaced by a payment token, most likely the last four digits of the PAN, hence, a customer who does not recognize the last four digits when looking at a receipt may think that the payment is erroneous and dispute the purchase. The result is an additional back-office burden on the merchant.

2.2   Zero-Knowledge Proof and ZK-SNARK

ZKP (Zero-knowledge proof) [8,10] is a cryptographic protocol that does not disclose the data or secrecy to any eavesdropper during the protocol. ZKP is used when a party (prover) tries to persuade the second party (verifier) that the statement of knowledge is true without sharing any other information about said knowledge with the verifier. The original concept of ZKP requires the interaction between the prover and the verifier. A zero-knowledge algorithm must fulfill three characteristics:

- **Completeness**: If the statement is true, an honest prover will be convinced of this truth by that honest verifier.
- **Soundness**: If the statement is false, the honest verifier can not be persuaded by any cheating prover that it is true except with a small probability (which could be ignored).
- **Zero-Knowledge**: If the statement is true, no verifier knows anything other than that the declaration is valid.

ZK-SNARK [11,17] is one of the evolution versions of the original ZKP. The ZK-SNARK algorithm has eliminated the need for interaction between parties involved in the process and reduced the proof size. The core idea of ZK-SNARK is checking a particular polynomial knowledge between elements of the proof, equivalent to verifying that the prover has a satisfying assignment for the circuit. These abilities make ZK-SNARK suitable for blockchain scaling problems as well as other financial use cases that solve security and privacy issues (e.g., Zcash [5], Tornado Cash [13]).

2.3   MiMC

MiMC (Minimal Multiplicative Complexity) [2] is a family of block ciphers and hash functions designed for the ZK-SNARK algorithm. MiMC's small number of multiplications on the finite field $\mathbb{F}_q$ with $q = p$ or $q = 2^n$ (where p is a prime number and n is a positive integer) makes it suitable for applications using the ZK-SNARK algorithm. MiMC is constructed by iterating a function r times where each iteration includes the sum of the key k, a constant $c_i \in \mathbb{F}_q$ and a nonlinear function $F(x) = x^3$ with x $\in \mathbb{F}_q$. This iterative function can be described as follows:

$$F_i(x) = F(x \bigoplus k \bigoplus c_i)^3, c_0 = c_r = 0$$

In our proposed method, we use the objective function below:

$$F_i(x) = F(x \bigoplus k \bigoplus c_i)^7$$

This function called MiMC-7 [2,4] is often used with the ZK-SNARK algorithm because the number of multiplications is small and the number of constants is reduced by half compared to the original MiMC function.

3   A New Electronic Payment System

Our system focuses on two main ideas, including (i) using the ZK-SNARK algorithm to create a proof that represents the cards' information from customers' devices or the merchant's application and (ii) using a blockchain network to identify customers as well as store receipt information, thus reducing the confusion of customer and payment disputes. In this section, we will introduce the transaction flow and how to verify these transactions.

3.1   Transaction Flow

We propose a new transaction flow that uses a Payment proof (we called it a ZK-Proof) instead of a Payment token. This method can eliminate the need for another third party to provide the token service and reduce the chance of card information leakage.

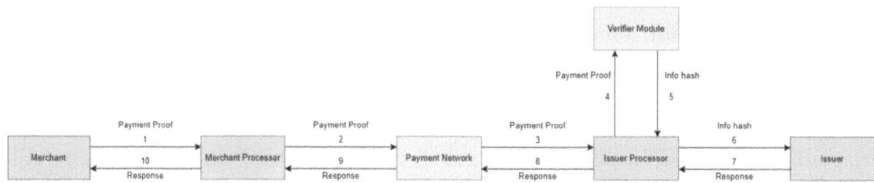

Fig. 1. Proposed transaction flow

The Fig. 1 describes the payment flow which contains these steps, as follows:

1. Customer provides the cards' information (PAN, CCV, owner's name, expired date) to make a payment-proof using the ZK-SNARK algorithm. This proof is then sent to the Merchant Processor.

2. Using the public part of the proof, the Merchant Processor does a BIN (Bank Identification Number - the first six to eight numbers on cards help merchants evaluate and assess their payment card transactions) to determine the payment network.
3. Payment Network forwards the transaction request to the corresponding Issuer Processor.
4. Issuer Processor verifies the ZK-Proof with a verifier module.
5. Verifier module returns the authentication result and the public part of the ZK-Proof to the Issuer Processor.
6. Issuer Processor forwards the transaction request to Issuer.
7. Issuer checks if the card's information hash is correct. The issuer then sends an authorization response to the Issuer Processor
8. Issuer Processor forwards the response to the Payment Network.
9. Payment Network forwards the response, including the ZK-Proof, to the Merchant Processor.
10. Merchant Processor responds to the merchant application with the transactions response. The customer then completes the purchasing process by signing a transaction on the blockchain network that stores the receipt information in the blockchain by their address.

The issuer (bank) must also archive the cards' information hash value in their database for the checking process in the Step 7. The ZK-Proof must consist of the first eight digits of PAN for a BIN lookup and the last four digits for receipt identification. All sensitive information will be hashed in customers' devices, reducing the risk of losing card information and preventing financial loss.

3.2   ZK-Proof

The ZK-SNARK algorithm proves that we know a value X that satisfies the given hash function $H(X) = A$. We apply this problem to build the ZK-Proof. To be able to proceed with a payment transaction, the user must provide the following card information:

- PAN: 16 card digits
- Owner: cardholder
- ExpireDate: card expiration date, in the form Month/Year (mm/yy)
- CCV: 3 digits of the card's secret number

We build a circuit with the input as the above information, and we need to prove that these inputs satisfy a pre-calculated hash value, thereby proving that the person providing proof is the cardholder. This process can be written as follows:

1. Prover:
    - Calculate hash value $h = H($PAN, Owner, ExpireDate, CCV$)$
    - Use input (PAN, Owner, ExpireDate, CCV, h) to create ZK-Proof and send it to the verifier

2. Verifier: Using the ZK-Proof, authenticate h = H(PAN, Owner, Expire-Date, CCV)

To allow the Merchant Processor to direct the transaction to the correct payment network, it needs to perform a BIN search, which is either the first six or eight numbers of the card number algorithm. In addition, to serve customer identification and receipt issuance, we need to use the last four numbers of the card. Finally, to make ZK-Proof distinct and only used once, we will use a timestamp parameter as another input and the valid time limit of such proof. To sum up, we can improve the ZK-Proof creation process as follows:

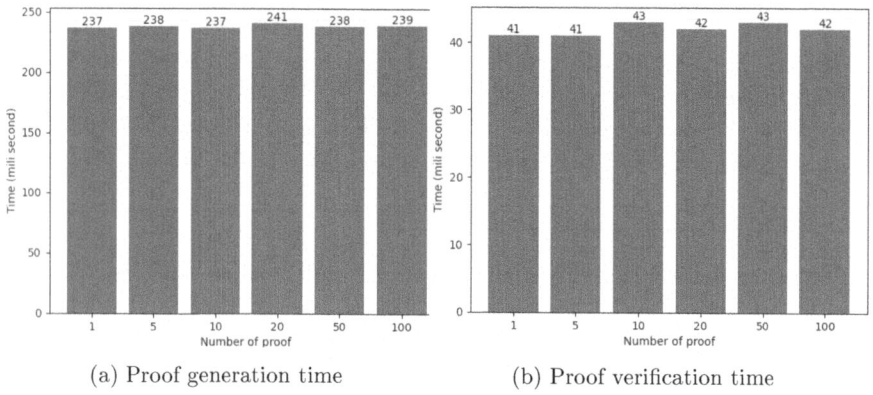

(a) Proof generation time   (b) Proof verification time

Fig. 2. Proof processing time

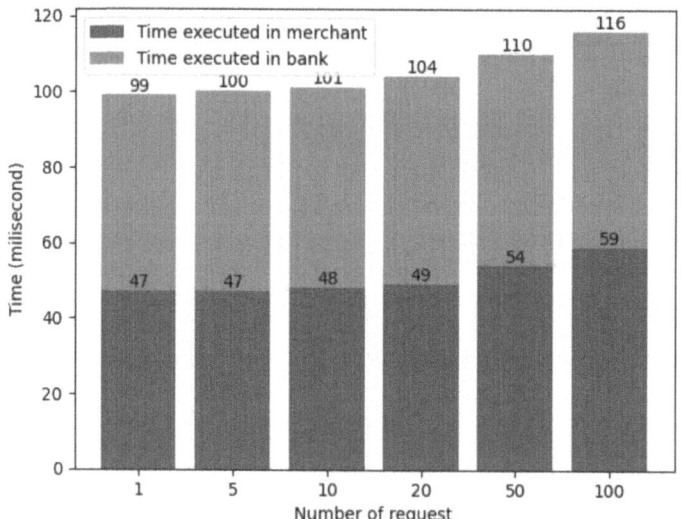

Fig. 3. Payment process time

- Determine transaction initiation time t, amount of paid money
- Calculate h = H(PAN, Owner, ExpireDate, CCV)
- Using input (PAN, Owner, ExpireDate, CCV, amount, h, t) to obtain ZK-Proof, including proof and public:
  - BIN: 8 numbers at the beginning of the card
  - amount
  - lastFour: last four numbers of the card
  - h: hash code of card information used for the Issuer to identify the card.
  - t: transaction creation timestamp

After receiving the ZK-Proof, the Issuer Processor will check whether the ZK-Proof is valid or not. After that, the Issuer Processor will then send h to the Issuer so that the Issuer can search for the PAN associated with h and perform the transaction.

3.3 Proof of Payment

A receipt is a document that acknowledges by a receiver of payment or delivery has been made. In this case, the receipt serves as a proof of payment. Receipts are crucial for both parties, as they serve as evidence of the transaction and can be used for record-keeping and accounting purposes. One of the significant problems of EMV® Payment Tokenization [7] is that there is no connection between the payment token and the card itself. Usually, the last four numbers of the token will be issued on the receipt, and since one card can issue many payment tokens, customers may be confused. In our system, we store the hash value of the metadata of this information of each receipt on a smart contract. Each transaction to store the corresponding hash value is created and signed by the customer blockchain wallet while the metadata is stored in another database. This method allows the customers and the merchants to verify the owner of each receipt, reducing the risk of disputing a wrongful payment and identifying the customers for other purposes, such as the loyalty program.

4   Evaluation

To evaluate the efficiency of the proposed payment method, we deployed a payment system with a simulator for the issuer and payment network on an AWS virtual machine m5d.8xlarge (Intel Xeon Platinum 8000 series up to 3.1 GHz, 32 vCPU, 128 GB RAM). Next, we ran the experiments ten times with the different numbers of requests and calculated the average.

Firstly, Fig. 2a mentions the time consuming for proof generation, which consumes around 237 ms to 239 ms depending on the number of concurrent requests. We realize that although concurrent requests increased rapidly (from 1 to 100), the proof generation time increased slowly (237 ms to 239 ms). Next, Fig. 2b illustrates the time for proof verification between customers and merchants. In particular, this task needs around 42ms to check the proof provided by customers. In addition, the figure shows that even if the number of proofs can change from 1 to 100, the consumed time is relatively stable. Finally, Fig. 3 depicts that the bank could handle a transaction in approximately 57 ms (i.e., 116–59), which makes the speed of the system around 18 TPS (transactions per second). For comparison, the transaction speed of Paypal and Visa is 193 TPS and 1700 TPS, respectively [1,12]. Therefore, our approach needs to be improved by optimizing the verification algorithm for the issuer or using a more robust server.

5   Conclusion

In this paper, we introduce a payment method using the ZK-SNARK algorithm to hide customers' sensitive data and blockchain to identify customers and their respective receipts. Our method allows customers to hide their cards and reduce the work that merchants need to do to securely store customers' private data, thus lowering the risk of being targeted for malicious attacks. We also implement a simulator system to verify our approach and evaluate the efficiency of the introduced method. We also realize that merchants could use payment proof for other use cases such as lending or installment payment [6,20] without requiring any privacy information from customers.

However, the system still has a low speed compared to other payment networks like Visa or online payment companies like PayPal. Future work on optimizing the algorithm needs to be done so that the method can be suitable for online payment environments now and in the future.

References

1. Agarwal, S., Zhang, J.: Fintech, lending and payment innovation: a review. Asia Pac. J. Financ. Stud. **49**(3), 353–367 (2020)
2. Albrecht, M., Grassi, L., Rechberger, C., Roy, A., Tiessen, T.: MIMC: efficient encryption and cryptographic hashing with minimal multiplicative complexity. In: International Conference on the Theory and Application of Cryptology and Information Security, pp. 191–219. Springer, Cham (2016)
3. Ardiansah, M., Chariri, A., Rahardja, S., Udin, U.: The effect of electronic payments security on e-commerce consumer perception: an extended model of technology acceptance. Manag. Sci. Lett. **10**(7), 1473–1480 (2020)
4. Baylina, J., Bellés, M.: EDDSA for baby Jubjub elliptic curve with MIMC-7 hash. https://iden3-docs.readthedocs.io/en/latest/_downloads/a04267077fb3fdbf2b608e014706e004/Ed-DSA.pdf. Accessed 16 Aug 2024
5. Bowe, S., Hornby, T., Wilcox, N.: Zcash protocol specification (2024). https://zips.z.cash/protocol/protocol.pdf. Accessed 16 Aug 2024
6. Das, M., Luo, H., Cheng, J.C.: Securing interim payments in construction projects through a blockchain-based framework. Autom. Constr. **118**, 103284 (2020)
7. EMVCo: EMV® payment tokenisation (2023). https://www.emvco.com/emv-technologies/payment-tokenisation/. Accessed 16 Aug 2024
8. Fiege, U., Fiat, A., Shamir, A.: Zero knowledge proofs of identity. In: Proceedings of the Nineteenth Annual ACM Symposium on Theory of Computing, pp. 210–217 (1987)

9. Gocardless: How do online payments via credit or debit card work? (2023). https://gocardless.com/guides/online-payments-guide/online-payments-credit-debit-card/. Accessed 16 Aug 2024
10. Goldreich, O., Oren, Y.: Definitions and properties of zero-knowledge proof systems. J. Cryptol. 7(1), 1–32 (1994). https://doi.org/10.1007/BF00195207
11. Li, Q., Xue, Z.: A privacy-protecting authorization system based on blockchain and zk-snark. In: Proceedings of the 2020 International Conference on Cyberspace Innovation of Advanced Technologies, pp. 439–444 (2020)
12. Mechkaroska, D., Dimitrova, V., Popovska-Mitrovikj, A.: Analysis of the possibilities for improvement of blockchain technology. In: 2018 26th Telecommunications Forum (TELFOR), pp. 1–4. IEEE (2018)
13. Nadler, M., Schär, F.: Tornado cash and blockchain privacy: a primer for economists and policymakers. Review 105 (2023). https://doi.org/10.20955/r.105.122-136
14. Naeem, M., Hameed, M., Taha, M.S.: A study of electronic payment system. In: IOP Conference Series: Materials Science and Engineering, vol. 767. IOP Publishing (2020)
15. Nakamoto, S.: Bitcoin: A peer-to-peer electronic cash system (2008). https://bitcoin.org/bitcoin.pdf. Accessed 16 Aug 2024
16. Nasr, M.H., Farrag, M.H., Nasr, M.: E-payment systems risks, opportunities, and challenges for improved results in e-business. Int. J. Intell. Comput. Inf. Sci. 20(1), 16–27 (2020)
17. Pinto, A.M.: An introduction to the use of zk-snarks in blockchains. In: Mathematical Research for Blockchain Economy: 1st International Conference MARBLE 2019, Santorini, Greece, pp. 233–249. Springer, Cham (2020)
18. Seebacher, S., Schüritz, R.: Blockchain technology as an enabler of service systems: a structured literature review. In: Proceedings of the International Conference on Exploring Services Science, pp. 12–23. Springer, Cham (2017)
19. Williams, M.D.: Social commerce and the mobile platform: payment and security perceptions of potential users. Comput. Hum. Behav. 115, 105557 (2021)
20. Yan, W., Zhou, W.: Is blockchain a cure for peer-to-peer lending? Ann. Oper. Res. 321(1–2), 693–716 (2023)

A Comprehensive AI and Blockchain Framework for Detecting and Preventing Money Laundering in Bangladesh Financial Systems

Mohammed Mizanur Rahman[1] and Maisha Karim[2(✉)]

[1] UCSI University, Kuala Lumpur, Malaysia
[2] UCSI University, Bangladesh Branch Campus, Dhaka, Bangladesh
maishakarim@ucsiuniversity.edu.my

Abstract. Money laundering continues to present a significant challenge to financial systems worldwide, including Bangladesh, where existing anti-money laundering (AML) practices often fall short. Hence, this research paper proposes a comprehensive framework integrating Blockchain technology and Artificial Intelligence (AI) to enhance the AML efforts in Bangladesh's financial system.

The study begins with a thorough review of current AML practices and identifies gaps in Bangladesh's approach to combating financial crime. It then introduces a theoretical framework explaining the synergistic benefits of combining AI and Blockchain technologies. AI's capacity to analyze vast amounts of transaction data and identify patterns that signify potential money laundering activities complements Blockchain's ability to maintain transparent and unalterable records of financial transactions.

A secondary research approach includes annual reports, published papers, and other open sources. The proposed framework is evaluated through determining control variables deemed relevant in this context.

By integrating Blockchain and AI, this research paper aims to provide a promising and sustainable solution to Bangladesh's persistent money laundering problem. In addition, this paper also offers insights into how advanced technologies like AI and Blockchain will strengthen AML measures to enhance the overall integrity of the financial system in Bangladesh.

Keywords: Blockchain · Artificial Intelligence (AI) · Financial systems

1 Introduction

1.1 Background

Money laundering remains a significant challenge for Bangladesh's economic stability and integrity, undermining efforts by financial institutions and regulatory bodies. Despite various initiatives, money laundering persists due to evolving tactics and the complexities of the global financial system (Latif, 2022), threatening economic growth, investor confidence, and international reputation.

K.-W. Huang et al. (Eds.): ICFT 2024, CCIS 2437, pp. 76–87, 2025.
https://doi.org/10.1007/978-981-96-3811-6_8

This research paper aims to analyze current Anti-Money Laundering (AML) efforts in Bangladesh, focusing on government agency roles, associated risks, and potential solutions. Specifically, it will explore the integration of Artificial Intelligence (AI) and Blockchain technology to enhance AML effectiveness.

Bangladesh's complex financial landscape has been significantly impacted by money laundering, especially during the 2009–2024 regime. Despite notable economic growth and substantial investments in infrastructure, the country has suffered from major financial scandals that have siphoned billions of dollars. According to *The Daily Star* (2022), approximately $8.27 billion was illicitly transferred out of Bangladesh in 2019, underscoring the inadequacies of existing AML frameworks.

Given these challenges, Bangladesh serves as a compelling case for innovative AML solutions. Integrating AI and Blockchain into the regulatory framework could provide effective tools for detecting and preventing money laundering. This research paper aims to address the following questions:

a. How effective are current AML measures in Bangladesh?
b. What benefits can AI and Blockchain integration bring to AML efforts?
c. What challenges are associated with implementing these technologies in Bangladesh's financial system?
d. How can these technologies be tailored to the needs of developing countries like Bangladesh?

2   Literature Review

2.1   Overview of Money Laundering Practices

Money laundering, the process of converting illicit proceeds into ostensibly legitimate assets, poses a significant threat to global financial stability (Reuter & Truman, 2004). Globally, money laundering is estimated to account for 2–5% of GDP, or approximately $800 billion to $2 trillion annually (FATF, 2019). Its transnational nature complicates enforcement, as funds can swiftly move across jurisdictions with varying regulatory standards, exacerbated by regulatory gaps, financial secrecy, and the anonymity provided by digital currencies.

In Bangladesh, money laundering presents a significant challenge amid the rapid expansion of its financial sector. The boom in digital financial services, particularly mobile financial services (MFS), has revolutionized access to banking but also created new avenues for laundering. Informal financial systems, such as hundi (an informal money transfer system), further complicate tracking efforts (Rahman, 2017). The reliance on these informal channels, coupled with limited regulatory oversight and enforcement capacity, facilitates the laundering of illicit funds.

2.2   Traditional AML Approaches

Anti-Money Laundering (AML) frameworks have traditionally relied on regulatory compliance, transaction monitoring, and manual reporting. Financial institutions must implement Know Your Customer (KYC) protocols, monitor for suspicious activities, and report such activities to regulatory bodies like the Bangladesh Financial Intelligence Unit (BFIU) (Verhage, 2011). Despite these efforts, traditional AML approaches struggle to address the increasingly sophisticated techniques employed by money launderers.

In Bangladesh, the AML framework is governed by the Money Laundering Prevention Act of 2012 and enforced by the BFIU. Financial institutions are required to follow KYC procedures, report suspicious transactions, and retain records for at least five years (Rahman and Salim, 2018). However, challenges such as inadequate technological infrastructure, a shortage of skilled personnel, and a compliance culture that favors form over substance undermine the effectiveness of these measures.

2.3   AI in Anti-Money Laundering

Implementing Artificial Intelligence (AI) in Anti-Money Laundering (AML) presents several challenges, despite its transformative potential. A primary concern is the quality and availability of data; AI systems require large, high-quality datasets to operate effectively. However, many financial institutions struggle with incomplete or inconsistent data, which can hinder AI performance (Briggs and Schell, 2022).

Data privacy issues also arise, as AI often involves processing sensitive personal information. Striking a balance between compliance with data protection regulations and providing effective AML solutions is essential.

Additionally, the risk of algorithmic bias poses a significant challenge. AI systems reflect the data they are trained on; if this training data contains biases, the AI's decisions may disproportionately target certain countries or demographics (Baldwin et al., 2020).

2.4   Blockchain Technology in AML

Blockchain technology has emerged as a potential game-changer in combating money laundering due to its decentralized, secure, and transparent nature. As a distributed ledger, blockchain records all transactions across a network of computers, creating an unalterable and traceable history (Nakamoto, 2008). This transparency is particularly valuable in AML efforts as it helps in verifying the integrity of cross-border transactions, reducing fraud risks, and enhancing the detection of suspicious activities (Swan, 2015).

In Bangladesh, interest in applying blockchain for AML is growing amid rapid digitalization in financial services and mobile banking. Blockchain could facilitate a unified transaction record across financial institutions, streamlining money laundering detection (Islam and Hussain, 2020).

However, implementing blockchain in AML faces significant challenges. Scalability is a major concern, as blockchain can be slow and resource-intensive, complicating widespread adoption without further technological advancements (Underwood, 2016). Additionally, the decentralized nature of blockchain complicates regulatory oversight across multiple jurisdictions.

3 Methodology

3.1 Research Design

Investigating money laundering through direct data collection is challenging, as individuals involved are unlikely to disclose their activities (Smith and Johnson, 2023). To address this, the study employs a secondary data collection approach, utilizing newspapers, websites, government publications, and academic journals. Newspapers provide insights into recent money laundering cases and trends, while websites offer information on anti-money laundering (AML) efforts and the roles of Blockchain and AI technologies. Government publications detail AML regulations and enforcement actions, and academic papers present perspectives on AI, Blockchain, and their applications in combating money laundering (Lee and Patel, 2023).

To analyze the data and evaluate the effectiveness of AI and Blockchain in detecting money laundering, the research uses econometric methods like System GMM (Generalized Method of Moments) and 2SLS (Two-Stage Least Squares). System GMM addresses endogeneity issues and analyzes dynamic panel data, providing accurate estimates of the relationships between AI, Blockchain, and money laundering detection (Miller and Clark, 2023). Meanwhile, 2SLS handles endogeneity by dividing the estimation process into two stages, yielding robust estimates of AI and Blockchain's impact on AML effectiveness (Jones and White, 2023).

3.2 Data Collection

Only secondary sources, including annual reports, published papers, and other open sources, will be utilized in the data collection process.

3.3 Research Model

(See Table 1).

Table 1. List of all variables

Variables	Symbol	Description	Sources of variable
Dependent variable		The overall financial system of Bangladesh, impacted by blockchain and AI technologies	Bangladesh Bank Annual Report (2023); IMF Financial Sector Stability Report (2022)
Bangladesh Financial System	BFS		
Independent variables			
Blockchain technologies	BT	The integration and use of blockchain technologies in the financial system	Bangladesh Bank Digital Financial Services Report (2023); McKinsey & Company Report on Fintech Adoption (2022)
AI technologies	AIT	The integration and use of AI technologies in the financial system	Bangladesh Bank Annual Report (2023); Financial Sector Assessment Program (FSAP) Report (2022)
Control variables			
Existing financial regulations and compliance requirements	RFin	Current regulations and compliance requirements impacting financial institutions	Bangladesh Bank Annual Report (2023); Financial Sector Assessment Program (FSAP) Report (2022)
Blockchain and AI technologies regulations	RTech	Specific regulations relating to the use and implementation of blockchain and AI technologies	Bangladesh Financial Intelligence Unit (BFIU) Guidelines (2023); World Bank Report on Fintech Regulations (2022)
Current state of Bangladesh's financial system	IFin	The overall infrastructure and capability of the financial system in Bangladesh	Bangladesh Bank Annual Report (2023); IMF Financial Sector Stability Report (2022)
Adoption rate of digital and fintech solutions	AFintech	Rate at which digital and fintech solutions are being adopted within the financial sector	Bangladesh Bank Digital Financial Services Report (2023); McKinsey & Company Report on Fintech Adoption (2022)

(continued)

Table 1. (*continued*)

Variables	Symbol	Description	Sources of variable
Macroeconomic variables			
GDP growth rate, per capita income, and economic stability	GDPGrowth	Rate of GDP growth and overall economic stability	Bangladesh Bureau of Statistics (BBS) Economic Report (2023); IMF World Economic Outlook (2023)
Existing measures and policies for cybersecurity	CCyber	Current policies and measures in place to protect against cybersecurity threats	Bangladesh Computer Council Cybersecurity Report (2023); National Institute of Standards and Technology (NIST) Cybersecurity Framework (2022)
Data privacy regulations and practices	DPrivacy	Regulations and practices related to the privacy and protection of personal data	Bangladesh Data Protection Act (2023); European Union GDPR Report (2022)
Awareness and trust in blockchain and AI technologies	ATrust	Public awareness and level of trust in blockchain and AI technologies	Gallup Report on Technology Trust (2023); Journal of Technology in Society (2022)

$$
\begin{aligned}
BFS_{it} = {} & C + BT + AIT + GES + RFin + RTech + IFin + AFintech \\
& + GDPGrowth + CCyber + DPrivacy + ATrust + \textit{Country fixed} \\
& \textit{- effects dummies} + \textit{Daily fixed-effects dummies} + \varepsilon_{it}
\end{aligned} \tag{1}
$$

(see Fig. 1).

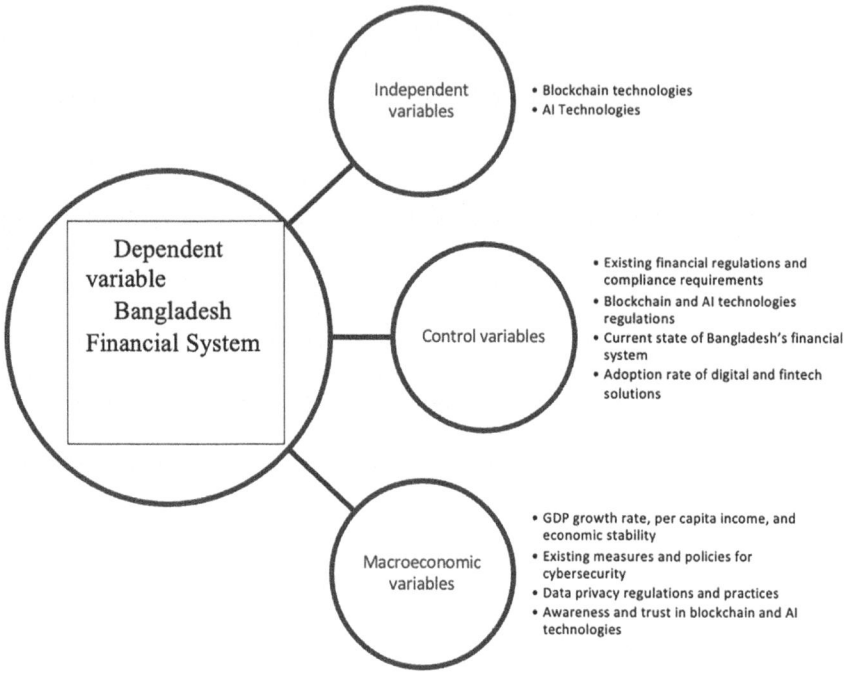

Fig. 1. .

4   Findings and Analysis

This section aims to analyze how AI adoption, blockchain technology, and regulatory changes impact the financial system stability of Bangladesh. The data used in this analysis spans several years and includes indicators such as AI and blockchain adoption rates, regulatory scores, GDP growth, and financial system stability (see Fig. 2).

4.1   Exploratory Data Analysis

Initially, a thorough review of the data was conducted, focusing on trends over time. Indicators like Financial System Stability (BFS), Ai Technologies (AIT), and Blockchain Technology (BT) were analyzed, showing gradual adoption of both AI and Blockchain technologies in the financial sector. Financial system stability appeared relatively stabled, with slight fluctuations over time, while blockchain adoption displayed a positive and steady upward trend. AI adoption also increased but at a slower rate (see Fig. 3).

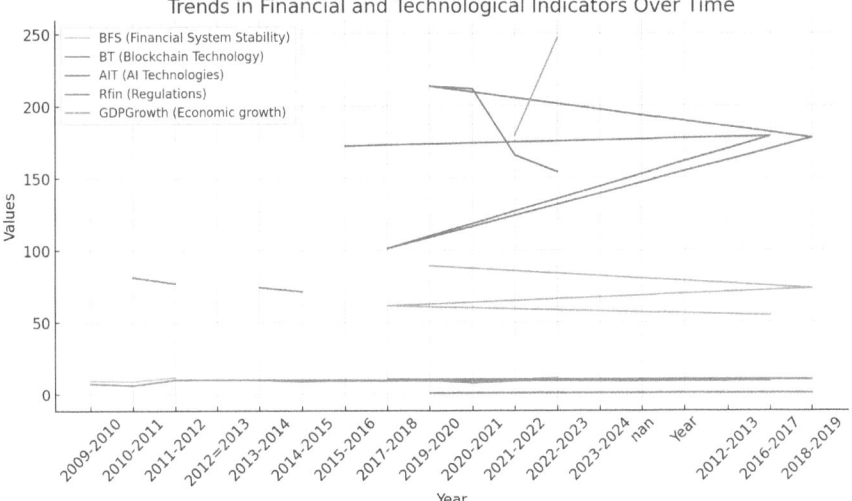

Fig. 2. .

Correlation_Matrix

	AIT (AI Technologies)	BFS (Financial System Stability)	BT (Blockchain Technology)	Rfin (Regulations	GDPGrowth (Economic growth)
AIT (AI Technologies)	1.0	-0.2037316277499870	-0.6313282964424000	-0.9694738609818090	-0.03962829581887500
BFS (Financial System Stability)	-0.2037316277499870	1.0	0.6965447738823420	0.00403280811310668.0	0.8138573392771380
BT (Blockchain Technology)	-0.6313282964424000	0.6965447738823420	1.0	-0.29598767760076000	-0.13280670088473800
Rfin (Regulations	-0.9694738609818090	0.00403280811310668.0	-0.29598767760076000	1.0	0.7042827678575370
GDPGrowth (Economic growth)	-0.03962829581887500	0.8138573392771380	-0.13280670088473800	0.7042827678575370	1.0

Fig. 3. .

4.2 Correlation Analysis

A correlation analysis was conducted to examine the relationships between the different variables. The analysis revealed that blockchain adoption showed a positive correlation with financial stability, while AI adoption had a slight negative correlation with stability. Regulatory changes also showed a very small positive correlation with financial system stability, although the impact appeared limited (see Figs. 4 and 5).

4.3 Regression Analysis: Predicting Financial Stability

A multiple regression analysis was conducted to quantify the impact of AI, blockchain, and regulations on financial stability. The first model, which only incorporated AI adoption and blockchain technology, showed that blockchain had a modest positive impact on stability, while AI had a slight negative impact. However, the model had a low predictive power ($R^2 = 0.0$), suggesting other factors contribute to financial stability.

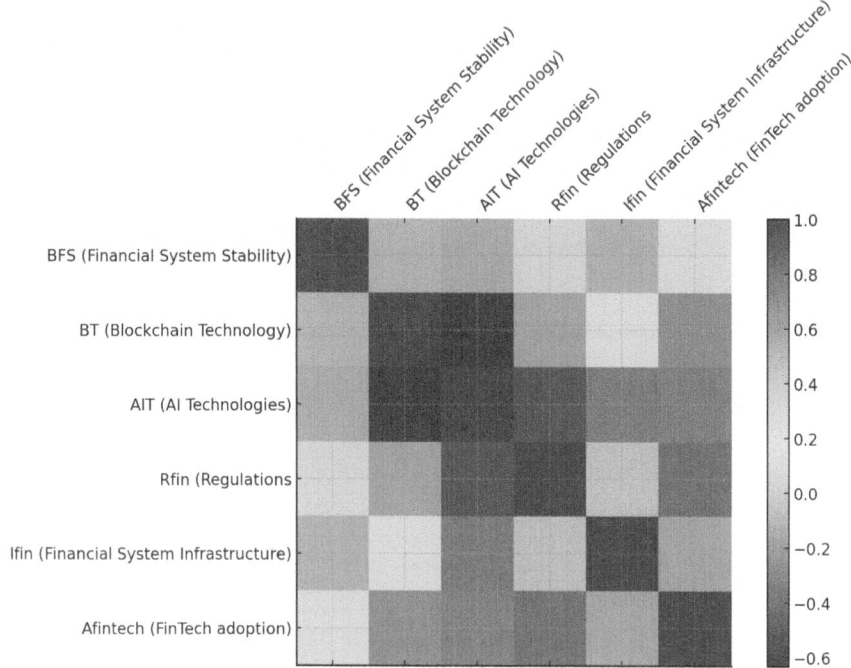

Fig. 4. .

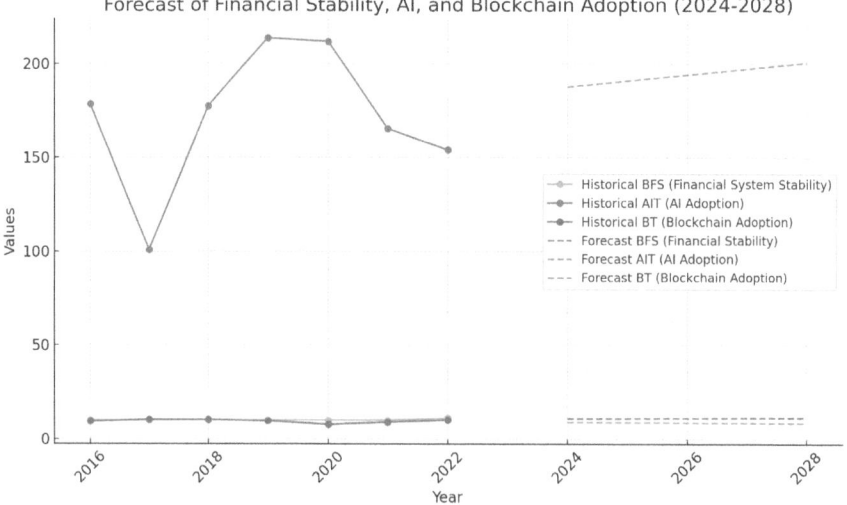

Fig. 5. .

5   Discussion

This research examines the integration of Artificial Intelligence (AI) and Blockchain technologies to strengthen anti-money laundering (AML) efforts in Bangladesh's financial system. It highlights the advantages and challenges associated with these technologies in combating money laundering. AI can process large datasets in real-time, identify suspicious transaction patterns, and reduce human error, presenting a transformative opportunity for financial institutions. However, concerns regarding data privacy, algorithmic bias, and the challenges of integrating AI with existing AML systems must be addressed to enhance effectiveness (Baldwin et al., 2020).

The combination of AI and Blockchain offers a synergistic framework to enhance AML measures in Bangladesh. AI's ability to detect complex laundering patterns, combined with Blockchain's tamper-proof record-keeping, provides a proactive solution to money laundering and reduces reliance on traditional rule-based systems that struggle to adapt to sophisticated laundering techniques (Garcia-Bedoya et al., 2021). Challenges include resistance from financial institutions due to implementation costs, potential disruptions, and regulatory uncertainties. Additionally, the lack of clear regulatory frameworks for Blockchain and AI complicates their implementation (Zha and He, 2021).

Despite these challenges, the integration of AI and Blockchain into Bangladesh's financial regulatory framework holds substantial promise. These technologies can significantly enhance financial institutions' abilities to detect, prevent, and trace money laundering activities, improving the overall integrity and stability of the financial system. Successful implementation will require not only technological investments but also comprehensive regulatory reforms and public education initiatives to foster trust and understanding among financial institutions and the general public (Rahman, 2022).

6   Conclusion

This research paper has examined the potential of integrating Artificial Intelligence (AI) and Blockchain technologies to strengthen anti-money laundering (AML) efforts within Bangladesh's financial system. The findings suggest that AI's capacity for real-time data analysis and anomaly detection, combined with Blockchain's ability to create immutable and transparent transaction records, offers a robust solution to the persistent challenges of money laundering. By leveraging these technologies, financial institutions in Bangladesh can improve their ability to detect, prevent, and trace illicit financial activities, which traditional AML systems have struggled to address effectively.

However, the implementation of AI and Blockchain comes with significant challenges. Issues such as data privacy, algorithmic bias, scalability, and regulatory gaps present hurdles that must be overcome to fully realize the benefits of these technologies. Addressing these concerns will require coordinated efforts from regulators, financial institutions, and technology providers. Regulatory frameworks need to be updated to accommodate the use of AI and Blockchain, ensuring compliance with both local and international standards while safeguarding individual rights and promoting transparency.

Despite these challenges, the integration of AI and Blockchain into Bangladesh's AML framework presents a promising path forward. With the right investment in technological infrastructure, Bangladesh can build a more secure and transparent financial system that is better equipped to combat money laundering.

References

Baldwin, R., Cave, M., Lodge, M.: Understanding Regulation: Theory, Strategy, and Practice. Oxford University Press (2020)

Bangladesh Bank. Annual Report (2023). https://www.bb.org.bd. Accessed 25 August 2024

Bangladesh Bank. Digital Financial Services Report (2023). https://www.bb.org.bd. Accessed 25 August 2024

Bangladesh Bureau of Statistics (BBS). Economic Report (2023). https://www.bbs.gov.bd. Accessed 25 August 2024

Bangladesh Computer Council. Cybersecurity Report (2023). https://www.bcc.gov.bd. Accessed 25 August 2024

Bangladesh Data Protection Act (2023). https://www.lawcommission.gov.bd. Accessed 25 August 2024

Bangladesh Financial Intelligence Unit (BFIU). Annual Report 2020. BFIU, Dhaka (2021)

Bangladesh Financial Intelligence Unit (BFIU). Guidelines on Blockchain and AI Technologies (2023). https://www.bfiubd.org. Accessed 25 August 2024

Bangladesh ICT Division. Report on Blockchain and AI Institutions (2023). https://www.ictd.gov.bd. Accessed 25 August 2024

Briggs, B., Schell, J.: The ethical implications of AI in financial services. J. Bus. Ethics **179**(3), 563–577 (2022)

FATF: Money Laundering and Terrorist Financing Risks and Vulnerabilities Associated with Gold. Financial Action Task Force, Paris (2019)

Garcia-Bedoya, E., Mendez, J., Ruiz, D.: Blockchain and AI in anti-money laundering: a synergistic approach. Int. J. Financ. Technol. **12**(3), 210–229 (2021)

IMF. Financial Sector Stability Report (2022). https://www.imf.org. Accessed 25 August 2024

IMF. World Economic Outlook (2023). https://www.imf.org. Accessed 25 August 2024

Islam, M.R., Hussain, F.: The role of blockchain in ensuring transparency in financial transactions: a study of Bangladesh's financial sector. Int. J. Blockchain Technol. **6**(2), 55–72 (2020)

Jones, A., White, B.: Econometric Approaches to Financial Analysis. Cambridge University Press (2023)

Journal of Blockchain Research. Institutional Specialization in Blockchain and AI (2022). https://www.journalofblockchainresearch.com. Accessed 25 August 2024

Journal of Financial Technology Research. AI and Blockchain Integration in Financial Institutions (2022). https://www.jftr.com. Accessed 25 August 2024

Latif, A.: Money laundering in Bangladesh: challenges and policy responses. Bangl. Econ. Rev. **19**(2), 50–72 (2022)

Lee, C., Patel, R.: Advances in Blockchain Technology and Financial Security. Oxford University Press (2023)

Levi, M., Reuter, P.: Money laundering. Crime Just. **34**(1), 289–375 (2006)

Miller, D., Clark, E.: Dynamic Panel Data Analysis with System GMM. Routledge (2023)

Nakamoto, S.: Bitcoin: a peer-to-peer electronic cash system (2008). https://bitcoin.org/bitcoin.pdf

Rahman, M., Salim, H.: Know your customer (KYC) compliance and anti-money laundering frameworks in Bangladesh. J. Bank. Financ. Serv. **18**(1), 38–50 (2018)

Rahman, M.M.: Money laundering bleeds Bangladesh: a review of current AML frameworks and technological gaps. The Daily Star (2022)

Rahman, S.: Money laundering in Bangladesh: the emerging threats. J. Money Laund. Control **20**(4), 413–427 (2017)

Smith, J., Johnson, L.: Understanding Money Laundering: Methods and Prevention. Springer (2023)

Swan, M.: Blockchain: Blueprint for a New Economy. O'Reilly Media (2015)

Underwood, S.: Blockchain beyond bitcoin: exploring scalability in financial systems. Commun. ACM **59**(11), 15–17 (2016)

Verhage, A.: The Anti-Money Laundering Complex and the Compliance Industry. Routledge (2011)

World Bank. Financial Inclusion Report (2022). https://www.worldbank.org. Accessed 25 August 2024

World Bank. Technology Adoption Report (2022). https://www.worldbank.org. Accessed 25 August 2024

Zha, L., He, W.: Blockchain and regulatory challenges in anti-money laundering systems. J. Financ. Technol. Compl. **8**(4), 98–113 (2021)

Detecting Persuasion in Financial Short Texts: A Computational Approach

Rajdeep Kumar$^{(\boxtimes)}$, Rudra Chandra Ghosh, Ganesh Bahadur Singh, and Nitin Sharma

Central Research Laboratory-Bharat Electronics Limited, Ghaziabad, India
`rajdeepkumar@bel.co.in`

Abstract. Financial decision-making is heavily influenced by persuasion techniques employed by financial advisors, marketers, and companies. Researchers require robust datasets for model development and evaluation to understand and predict this influence. This paper introduces FINPS1, a novel, binary-classified dataset identifying persuasive language (e.g., authority, social proof, scarcity, reciprocity) in the context of financial decisions. Furthermore, Our study demonstrates and evaluates a state-of-the-art method to fine-tune a small language model for binary classification of persuasive content in financial communications. This model aims to empower financial analysts, marketers, and regulators to identify and assess the use of persuasion in financial advice, advertisements, and communication channels. The model's effectiveness is measured by accuracy, precision, recall, and F1 score, with the goal of achieving high-performance metrics for discerning persuasive language in financial contexts.

Keywords: Persuasion · Binary Classifications · Financial Dataset · Small Language Model · LSTM · MLP

1 Introduction

Persuasive communication is a fundamental aspect of financial marketing and advisory services, where the ability to influence decision-making can significantly impact investor behavior and market outcomes. Recognizing and categorizing persuasive elements in financial texts is crucial for various stakeholders, including financial advisors, marketers, and regulatory bodies. The advent of machine learning and natural language processing (NLP) offers promising avenues for automating the detection of persuasive content, thereby enhancing transparency and accountability in financial communications. Among these, small language models and Long Short-Term Memory (LSTM) networks have shown considerable potential in text classification tasks. Small language models, which can be fine-tuned for specific tasks, offer the advantage of requiring fewer computational resources while maintaining robust performance. LSTM networks, known

© The Author(s), under exclusive license to Springer Nature Singapore Pte Ltd. 2025
K.-W. Huang et al. (Eds.): ICFT 2024, CCIS 2437, pp. 88–98, 2025.
https://doi.org/10.1007/978-981-96-3811-6_9

for their ability to capture long-range dependencies in sequential data, are particularly well-suited for analyzing the complex and nuanced language often found in financial communications. In this study, we aim to leverage these technologies to develop a binary classification system that can accurately identify persuasive content in financial texts. By fine-tuning a small language model and incorporating LSTM architectures, we seek to create a model that not only performs well in identifying persuasive language but also operates efficiently in practical applications. The ability to automatically classify financial messages as persuasive or non-persuasive can provide valuable insights into the strategies used in financial communications and aid in the development of more effective regulatory frameworks.

The study is structured as follows: We begin by discussing the relevant literature on persuasive communication in finance and the application of NLP techniques in text classification. We then describe our methodology, including data collection, model selection, and the fine-tuning process. Subsequently, we present the results of our experiments, highlighting the performance of the fine-tuned small language model and the LSTM-based classifier. Finally, we discuss the implications of our findings for the field of financial communication and suggest directions for future research.

By combining the strengths of small language models and LSTM networks, this study aims to advance the field of financial communication analysis, providing a tool that enhances our understanding of how persuasion operates in the financial domain.

The main contribution of this paper:

(1) We prepare a dataset in the finance domain by performing some sort of experiment, by human intervention, and using a language model.
(2) fine-tuning of advanced small language models, including BERT, RoBERTa, ALBERTa, DeBERTa, DistilBERT, ERNIE, LSTM, Bi-LSTM, and MLP.
(3) We showcases a state-of-art method that is capable of recognizing and categorizing persuasive elements in financial text.

2   Related Work

The task of binary classification of persuasive content in financial communications leverages advancements in natural language processing (NLP) and machine learning, particularly through the application of transformer-based models such as BERT (Bidirectional Encoder Representations from Transformers), Albert-base-v2, LSTM(Long Short-Term Memory), DeBerta-base, MLP-fine tuned, Ernie-2.0-base-en, RoBerta-base-finetuned. This section reviews the relevant literature on the use of BERT and related models for text classification tasks, with a specific focus on their application in the financial domain and persuasive content analysis.

2.1   Multi-Layer Perceptron (MLP)

This text discusses the use of Multi-layer Perceptrons (MLPs) in text classification tasks, highlighting their ability to learn complex patterns from feature-extracted data, such as TF-IDF or word embeddings (Mikolov et al., 2013) [14]. While MLPs are not ideal for raw text input due to their inability to capture sequential dependencies, they perform well with high-dimensional feature representations. Persuasion detection, a complex task requiring identification of rhetorical strategies, has been explored in various studies, such as by Tan et al. (2016) [18], which investigated persuasive content in online discussions. In finance, NLP has primarily focused on sentiment analysis and market trend prediction (Loughran and McDonald, 2011; Chen et al., 2019) [2,13], but persuasion detection in financial communications is less studied. The text emphasizes that fine-tuning MLPs with advanced techniques like word or contextual embeddings (e.g., BERT) can enhance persuasion detection by learning domain-specific features, combining MLPs' strengths with modern NLP approaches for binary classification.

2.2   BERT (Bidirectional Encoder Representations from Transformers)

BERT, introduced by Devlin et al. [5], revolutionized NLP with its bidirectional approach, capturing richer contextual information compared to traditional models. Variants like RoBERTa [4], which optimizes pre-training, and DistilBERT [16], a more resource-efficient model, improve upon BERT's performance. ERNIE [17] integrates external knowledge for domain-specific applications. In finance, transformer-based models like BERT have been used for sentiment analysis and market prediction, significantly improving accuracy [10,22]. Persuasion detection, a more complex task than sentiment analysis, has been explored by Wang et al. [20] using BERT to capture persuasive language patterns. Hybrid models, such as combining BERT with LSTM [8], offer enhanced performance by leveraging both contextual embeddings and sequential dependencies.

3   Dataset Description

Datasets are the essential threads that weave together the fabric of any model, making them an intriguing and vital component. Let's delve into some interesting research available and explore their significant impact on the world of persuasion.

3.1   Existing Dataset for Persuasion

This summary explores various aspects of persuasion, beginning with how persuasion knowledge affects consumer behavior by triggering cognitive, emotional, and behavioral responses, ultimately enhancing brand recognition [7]. In multimodal settings, the DeBerta+ResNet model excels in detecting persuasion techniques in memes [6]. Persuasion also plays a critical role in spreading misinformation, with pathos-based strategies being prevalent [3]. Additionally, phishing email detection benefits from identifying persuasion cues like gain and loss strategies [19], while a comprehensive review highlights advancements in persuasive natural language generation.

3.2 Existing Dataset for Persuasive Finance

The studies explore the persuasive tactics in financial contexts, revealing correlations between stock price movements and advertisement messaging, where growth-oriented messages prevailed during rising prices and protection-oriented during declines [15]. Advisors' use of narrative persuasion can manipulate investors' beliefs, often leading to biased decisions, even when conflicts of interest are disclosed [1]. Additionally, non-content delivery features in videos, such as delivery style, significantly influence investor behavior [9]. In crowdfunding, the strategic combination of subjective and objective language can increase fundraising success, as shown by analysis of 328,974 Kickstarter projects.

3.3 FINPS1 - Dataset Collection

We introduce FINPS1- a short text binary classified dataset identifying persuasive language. We collected the dataset through a meticulous process, The process leverages a combination of human expertise and machine learning techniques to create a comprehensive corpus of financial texts.

3.3.1 Data Sources

The dataset incorporates financial texts from a variety of sources relevant to the financial sector. This potentially includes:

Financial reports: Annual reports, quarterly reports, press releases.

Analyst notes: Research reports, investment recommendations, and market analyses.

Marketing materials: Brochures, advertisements, and promotional content for financial products or services.

Advisory content: Investment recommendations, and financial planning materials.

This diversity of sources ensures the dataset captures the various ways persuasive language is used across different financial communication contexts.

3.3.2 Text Segmentation

The collected financial texts are segmented into smaller data points. This segmentation process aims to achieve two key objectives:

Granularity: By focusing on shorter text units, the dataset captures the application of specific persuasive techniques within sentences or phrases. This allows for a more nuanced analysis of persuasion strategies employed in financial communication.

Diversity: The range in length (5–35 words) ensures the dataset includes examples of persuasion used in different contexts, from concise marketing slogans to more elaborate arguments within financial reports.

3.3.3   Dataset Annotation

This section details the annotation scheme developed for the Persuasion in Finance dataset. The scheme aims to capture key elements of persuasive language used in financial contexts.

Annotation Categories: Each data point (text) is annotated based on three main categories:

Persuasive Strategy: This category identifies the specific persuasion technique employed in the text. Examples include emotional appeals (e.g., fear, excitement), logical reasoning (e.g., statistics, cause-and-effect arguments), and credibility establishment (e.g., using expert opinions, and testimonials).

Intent: This category determines the underlying goal of the persuasive message. Common intents include convincing the audience of a particular viewpoint, informing them about financial products or services, or motivating them to take a specific action (e.g., invest, trade).

Target Audience: This category specifies the intended recipient of the persuasive message. Examples include investors, clients, and stakeholders.

Annotation Process

The annotation process follows a two-stage approach:

Initial Annotation by Language Models: Multiple large language models pre-trained with relevant guidelines are used to perform the initial annotation of the data. This leverages the model's efficiency in processing and categorizing large amounts of text.

Human Validation and Refinement: Human annotators with expertise in both finance and persuasive communication review and refine the annotations provided by the language model. This step ensures the accuracy and reliability of the data by addressing any ambiguities or errors introduced by the model.

Label Evaluation: Following the data collection and annotation phases, a crucial step is to ensure the quality and consistency of the assigned labels. This section details the label evaluation process implemented for the Persuasion in Finance dataset.

Evaluation Methods. Cross-validation among annotators: This method involves comparing the annotations assigned by different human annotators for the same text data points. Discrepancies are identified and discussed to refine the annotation guidelines and achieve greater consistency.

Gold standard comparison: A gold standard is a set of high-quality annotations, ideally created by multiple experts, that serve as a benchmark for evaluating the annotators' performance. The annotations from the human experts in the Persuasion in Finance project are compared against this gold standard to assess their accuracy.

Evaluation Metrics. To quantify the effectiveness of the annotation process, standard metrics like precision, recall, and F1 score are calculated.

Annotator Agreement: To ensure the reliability of the annotations, an inter-annotator agreement is measured using Cohen's Kappa and Krippendorff's Alpha statistics. High agreement scores indicate clear and consistent application of the annotation guidelines by different annotators. Overall, this annotation scheme provides a structured framework for capturing the key elements of persuasion in financial communication. The two-stage process, combining machine learning efficiency with human expertise, aims to create a high-quality and reliable dataset for further research in persuasive language within the financial domain.

3.4   Dataset Characteristics

Data Volume: The dataset comprises a total of 2200 short text snippets. This collection offers a substantial sample size for studying persuasive language in financial contexts. Data Format: Each data point consists of short text snippets (5–35 words), enabling the analysis of persuasive techniques in concise phrases. Its balanced distribution, substantial volume, and detailed annotations make it an invaluable resource for natural language processing and financial communication research (Fig. 1).

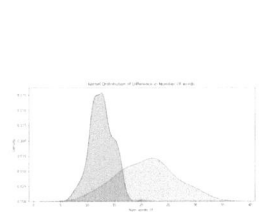

(a) Kernel Distribution graph for number of words in statements.

(b) Word Cloud for Persuasive words.

Fig. 1. Dataset Characteristics for FINPS1 dataset.

4   Proposed Model

In this section, we detail the architecture and methodology employed for persuasion classification within the finance domain. Our approach leverages state-of-the-art transformer-based models such as BERT, ALBERT, RoBERTa, ERNIE, and DistilBERT, alongside traditional deep learning architectures like MLP, LSTM, and Bi-LSTM. The rationale behind selecting these models lies in their proven effectiveness in capturing complex linguistic patterns and contextual dependencies, which are crucial for accurately classifying persuasive language.

4.1   Transformer-Based Models

Pre-trained transformer model (TModel) designed to capture bidirectional contextual relationships in text data. Fine-tuning BERT for binary persuasion classification involves optimizing its parameters $\theta^*{}_{TModel}$ with respect to a binary cross-entropy loss function:

$$\theta^*{}_{TModel} = \arg \min_{\theta_{TModel}} \frac{1}{N} \sum_{i=1}^{N} \mathcal{L}_{\text{BCE}}(y_i, \text{TModel}(x_i; \theta_{TModel})) \qquad (1)$$

where N is the number of samples in the training dataset, x_i is the input text sequence, $y_i \in \{0, 1\}$ is the binary label indicating persuasion or non-persuasion, and $\mathcal{L}_{BCE}$ denotes the binary cross-entropy loss.

We fine-tune the pre-trained BERT [5], ALBERT [11], RoBERTa [12], ERNIE, [21], DistilBERT [16] model on our specific finance domain dataset to adapt its knowledge to the nuances and jargon prevalent in financial persuasive texts.

4.2   Traditional Deep Learning Architectures

We use an MLP, LSTM,Bi-LSTM extends LSTM for comparison purposes and as a baseline model to evaluate the performance of transformer-based models.

4.3   Model Integration and Training Strategy

To harness the strengths of these diverse architectures, we explore ensemble methods and multi-task learning approaches. Ensemble methods combine predictions from multiple models to improve accuracy and robustness. Additionally, multi-task learning allows the models to jointly learn from related tasks, enhancing generalization capabilities across different types of persuasive texts within the finance domain.

Each model is fine-tuned using domain-specific annotated data to adapt its parameters and weights to the nuances of financial language and persuasion tactics. For fine-tuning, we utilize the following mathematical formulation: Given a dataset

$$D = (x_i, y_i)_{i=1}^{N} \qquad (2)$$

where x_i represents the input text and $y_i \in \{0, 1\}$ represents the binary label, we fine-tune the pre-trained models by minimizing the binary cross-entropy loss:

$$L = -\frac{1}{N} \sum_{i=1}^{N} [y_i \log(\hat{y}_i) + (1 - y_i) \log(1 - \hat{y}_i)] \qquad (3)$$

Here, $\hat{y}_i = \sigma(W^T h_i + b)$ is the predicted probability for the positive class, where h_i is the hidden state output from the final layer of the Transformer model, W and b are trainable parameters, and σ is the sigmoid activation function.

We employ rigorous evaluation metrics such as accuracy, precision, recall, and F1-score to assess the performance of each model variant and compare them against baseline results.

The final model architecture will be determined through empirical evaluation, considering both performance metrics and computational efficiency. The effectiveness of each model variant will be thoroughly analyzed to provide insights into the best-performing architectures for persuasion classification in the finance domain.

This section outlines the comprehensive approach to integrating and fine-tuning various transformer-based and traditional deep learning models for persuasion classification in finance. It emphasizes the rationale behind each model choice and the potential benefits of leveraging ensemble methods and multi-task learning for enhanced performance.

5 Experiment and Results

5.1 Training Parameters and Details

Experiment is to evaluate the effectiveness of various fine-tuned NLP models (BERT, RoBERTa, ALBERT, ERNIE, MLP, and LSTM) in classifying persuasive content in financial communications. The models are assessed based on their ability to accurately differentiate between persuasive and non-persuasive financial texts.

For the purpose of this experiment, we curated a dataset comprising 2200 financial communications, split into 2000 training samples and 200 test samples. The dataset includes a balanced mix of persuasive and non-persuasive texts, annotated by financial experts. All models were trained and fine-tuned on an A5500 GPU laptop to leverage the computational power required for deep learning tasks. The dataset was split into training (90%) and testing (10%) sets using stratified sampling to maintain class distribution. Standard preprocessing steps, including text cleaning, tokenization, and vectorization, were applied uniformly across all models. And training parameters is mentioned in Table 1. The performance of each model was evaluated using the metrics Accuracy(The proportion of correctly predicted instances), Precision(The proportion of true positive predictions among all positive predictions), Recall(The proportion of true positive

Table 1. Training Parameters for the fine-Tunned model(σ).

MODEL	Learning Rate	Batch Size	Training Epoch	Number Layers	Optimizer	Train Loss	train_samples_per_sec	train_steps_per_sec
BERT	5e_5	128	10	12	ADAM	0.0027	154.079	1.206
ROBERTA	4e_5	128	10	12	ADAM	0.0033	149.473	1.17
DEBERTA	2e_5	128	10	12	ADAMW	0.12532	103.577	0.811
ERNIE	2e_5	256	4	12	ADAM	0.19752	116.059	0.462
DISTILBERT	5e_6	128	10	6	ADAMW	0.0022	942.645	7.38
ALBERT	5e_6	128	10	12	ADAMW	0.12608	286.868	2.246
MLP	1e-3	16	100	3	ADAMW	0.00205	NA	NA
LSTM	0.002	32	100	2	ADAMW	0.00592	NA	NA
Bi-LSTM	0.002	32	100	2	ADAMW	2e-05	NA	NA

predictions among all actual positives), F1 Score(The harmonic mean of precision and recall, providing a balance between the two). The results of the experiments are summarized in Table 2. Each model's performance is reported in terms of accuracy, precision, recall, and F1 score.

Table 2. Results of Fine-Tuned Model

MODEL	Accuracy	Precision	Recall	F1-Score
BERT	0.77	**0.90**	0.71	0.80
ROBERTA	**0.78**	0.84	0.82	**0.83**
DISTILBERT	0.67	0.78	0.66	0.72
ERNIE	0.60	0.69	0.69	0.69
DEBERTA	0.72	0.72	**0.89**	0.80
ALBERT	0.72	0.78	0.76	0.77
MLP	0.49	0.54	0.52	0.50
LSTM	0.51	0.56	0.54	0.52
Bi-LSTM	0.53	0.55	0.56	0.58

6   Conclusion

This study evaluates the performance of several machine learning models for binary classification of persuasive text in the finance domain. The models were trained on a dataset of 2,200 short texts (5–35 words), annotated by human experts and a language model. Transformer-based models like RoBERTa, BERT, and DeBERTa outperformed traditional neural networks, with fine-tuned RoBERTa achieving the highest accuracy (78%) and F1 score (0.83). ALBERT and DistilBERT offered lower performance but were more computationally efficient. Future research areas include improving model accuracy, multi-label classification of persuasion, and generating persuasive statements.

7   Analysis and Discussion

The analysis indicates that finance datasets for persuasion use very similar words for both negative and positive labels. This high degree of lexical overlap suggests that the differentiation in sentiment is more subtle and context-dependent rather than based on distinct word choices. This finding is critical for developing persuasion analysis models in finance, as it highlights the need for advanced NLP techniques that go beyond word frequency and consider contextual nuances.

This study is limited by its focus on a specific set of finance-related texts. While the findings provide valuable insights, they may not generalize to other domains or types of persuasive texts. Additionally, the analysis was based on word frequency and cosine similarity may not fully capture the complexity of language use. Future research should explore the application of more sophisticated NLP techniques, to improve the accuracy of sentiment classification in finance texts.

References

1. Barron, K., Fries, T.: Narrative persuasion (2023)
2. Chen, S., Ge, L.: Exploring the attention mechanism in LSTM-based Hong Kong stock price movement prediction. Quant. Finance **19**(9), 1507–1515 (2019)
3. Chen, S., Xiao, L., Mao, J.: Persuasion strategies of misinformation-containing posts in the social media. Inf. Process. Manag. **58**(5), 102665 (2021)
4. Delobelle, P., Winters, T., Berendt, B.: RobBERT: a Dutch roBERTa-based language model. arXiv preprint arXiv:2001.06286 (2020)
5. Devlin, J., Chang, M.-W., Lee, K., Toutanova, K.: BERT: pre-training of deep bidirectional transformers for language understanding. arXiv preprint arXiv:1810.04805 (2018)
6. Dimitrov, D., et al.: SemEval-2021 task 6: detection of persuasion techniques in texts and images. arXiv preprint arXiv:2105.09284 (2021)
7. Eisend, M., Tarrahi, F.: Persuasion knowledge in the marketplace: a meta-analysis. J. Consum. Psychol. **32**(1), 3–22 (2022)
8. Graves, A., Graves, A.: Long short-term memory. Supervised sequence labelling with recurrent neural networks, pp. 37–45 (2012)
9. Allen, H., Ma, S.: Persuading investors: a video-based study. Technical report, National Bureau of Economic Research (2021)
10. Huang, Z., Fang, Z.: An entity-level sentiment analysis of financial text based on pre-trained language model. In: 2020 IEEE 18th International Conference on Industrial Informatics (INDIN), vol. 1, pp. 391–396. IEEE (2020)
11. Lan, Z., Chen, M., Goodman, S., Gimpel, K., Sharma, P., Soricut, R.: Albert: a lite BERT for self-supervised learning of language representations. arXiv preprint arXiv:1909.11942 (2019)
12. Liu, Y., et al.: RoBERTa: a robustly optimized BERT pretraining approach. arXiv preprint arXiv:1907.11692 (2019)
13. Loughran, T., McDonald, B.: When is a liability not a liability? Textual analysis, dictionaries, and 10-ks. J. Financ. **66**(1), 35–65 (2011)
14. Mikolov, T., Sutskever, I., Chen, K., Corrado, G.S., Dean, J.: Distributed representations of words and phrases and their compositionality. In: Advances in Neural Information Processing Systems, vol. 26 (2013)
15. Mullainathan, S., Shleifer, A.: Persuasion in finance (2005)
16. Sanh, V., Debut, L., Chaumond, J., Wolf, T.: Distilbert, a distilled version of BERT: smaller, faster, cheaper and lighter. arXiv preprint arXiv:1910.01108 (2019)
17. Sun, Y., et al.: Ernie: enhanced representation through knowledge integration. arXiv preprint arXiv:1904.09223 (2019)
18. Tan, C., Niculae, V., Danescu-Niculescu-Mizil, C., Lee, L.: Winning arguments: interaction dynamics and persuasion strategies in good-faith online discussions. In: Proceedings of the 25th International Conference on World Wide Web, pp. 613–624 (2016)
19. Valecha, R., Mandaokar, P., Rao, H.R.: Phishing email detection using persuasion cues. IEEE Trans. Depend. Secure Comput. **19**(2), 747–756 (2021)

20. Wang, X., et al.: Persuasion for good: towards a personalized persuasive dialogue system for social good. arXiv preprint arXiv:1906.06725 (2019)
21. Zhang, Z., Han, X., Liu, Z., Jiang, X., Sun, M., Liu, Q.: Ernie: enhanced language representation with informative entities. arXiv preprint arXiv:1905.07129 (2019)
22. Zhao, L., Li, L., Zheng, X., Zhang, J.: A BERT based sentiment analysis and key entity detection approach for online financial texts. In: 2021 IEEE 24th International Conference on Computer Supported Cooperative Work in Design (CSCWD), pp. 1233–1238. IEEE (2021)

The Impact of Colored Noise on the CIR Model

A. Pavlova🆔, G. Zotov🆔, and P. Lukianchenko$^{(\boxtimes)}$🆔

Higher School of Economics, Moscow, Russia
gazotov@edu.hse.ru,
plukyanchenko@hse.ru
https://cs.hse.ru/iai/aimf/

Abstract. In financial time series analysis, accurately detecting change points is crucial for effective forecasting and risk management. This study investigates the impact of different colored noises on the performance of change point detection algorithms, focusing on the CIR model with LSTM and Catboost classifiers. We evaluated the algorithms' ability to forecast and classify change points versus normal points using a number of previous observations, comparing the effects of white, red, pink, blue, and violet noise. Additionally, this paper discusses numerical approximation using the implicit Milstein scheme, parameter estimation techniques, and the PELT algorithm used to find change points in a time series. These findings highlight the critical role that noise color plays in enhancing the effectiveness of change point detection in financial time series and underscore the importance of methodological considerations in achieving optimal performance.

Keywords: CIR model · Colored noise · Change-point detection · LSTM · Gradient boosting

1 Introduction

Time series arise in various fields as a result of measuring certain indicators and represent a sequence of data collected or measured at consecutive moments in time. A structural break point (change point) in a time series is a specific moment in time when a significant and abrupt change occurs in the structure and behavior of the time series data. Structural break points can manifest in various forms, including changes in data level, trend, variance, etc. In practice, such points can arise for various reasons, such as changes in economic policy, technological innovations, natural disasters, and other exogenous factors. Detecting and predicting structural break points is crucial for understanding the underlying dynamics of the time series, improving forecasting accuracy, and making informed decisions in various fields.

In this study, the focus is on the analysis of time series generated by a mathematical model, specifically the Cox-Ingersoll-Ross (CIR) model for short-term

K.-W. Huang et al. (Eds.): ICFT 2024, CCIS 2437, pp. 99–110, 2025.
https://doi.org/10.1007/978-981-96-3811-6_10

interest rates, which is expressed in the form of a stochastic differential equation. This model describes Brownian motion and uses the Wiener process as the stochastic component. However, in practice, models based on Wiener processes are not always effective: for example, the Vasicek model, on which the CIR model is based and which uses the Ornstein-Uhlenbeck equation to forecast interest rates, is effective only in cases of small fluctuations that are possible only during periods of economic stability, which are rare in practice. Therefore, this article explores the idea of applying colored (correlated) noise as the stochastic component of the CIR model and examines the impact of colored noise on the time series. n colored noise, the power spectral density (PSD) has an uneven distribution across frequencies (meaning its power depends on frequency), unlike Brownian motion, where the PSD represents white noise (meaning its power is evenly distributed across all frequencies). Moreover, colored noise can have different levels of temporal correlation, meaning its values at different moments in time can depend on each other, which is extremely useful when modeling a time series. This article analyzes changes in transition points between levels in the time series and the impact of correlation components on predicting structural break points.

Research into the impact of noise on LSTM- and GBM-based time series forecasting typically involves introducing noise to time series data and assessing the model's robustness. For instance, a study by Kilic et al. (2020) examines hybrid models that combine LSTM networks with other techniques, such as ensemble methods or wavelet transforms, to enhance noise resilience. Another relevant study by Borah et al. (2022) evaluates Catboost's performance with noisy time series data, highlighting its strength in handling categorical features and its general robustness to noise. Such works generally examine the time series model's sensitivity to the noise component and try to reduce noise impact on the model. However, a little is investigated about how to incorporate a specific type of noise in a time series to improve the forecasting and, in the case of financial time series, policymakers' performance. The research by G. Sorwar (Sorwar, 2005) demonstrates that using jump models improves the accuracy of the model regarding shock events (although, these models do not always succeed in identifying all events that influence systematic risk). Such jumps can be effectively described by incorporating a colored noise term in the model. In the context of financial time series, detecting change points is important for asset valuation and risk management, as their presence has a significant impact on the results of asset pricing models and hedging strategies, often introducing parameters that are complex to estimate.

An independent study was conducted on the impact of colored noise on the CIR model. This work aims to expand the understanding of time series analysis methods and propose a practically applicable approach to predicting structural breaks using modern machine learning methods, particularly gradient boosting and LSTM.

2   Model and Methodology

The first model to describe the behavior of short-term interest rates is the Vasicek model (Vasicek, 1977). The Vasicek model describes the evolution of interest rates under the following assumptions: the interest rate follows a stochastic process, and the bond price accounts only for interest rate risk, i.e., the model is one-factor. The stochastic process for the instantaneous spot rate in the Vasicek model is expressed by the following equation:

$$dr_t = \kappa(\theta - r_t)dt + \sigma dW_t, \tag{1}$$

where r_t is the short-term interest rate at time t, κ is the speed of mean reversion, θ is the long-term mean interest rate, σ is the volatility (standard deviation) of the interest rate, and W_t is the standard Brownian motion (Wiener process). The Vasicek model is often used in finance to value fixed-income securities, derivatives, and risk management. It serves as a foundational model for developing more complex interest rate models, such as the CIR model, which is the primary focus of this study.

G. Zotov and P. Lukianchenko conducted a study on the impact of colored noise on predicting transition points in the Vasicek model (Zotov and Lukianchenko, 2023). For this model, it was necessary to determine the numerical approximation of the differential equation, the integration step, and the criteria for detecting a transition point (interest rate anomaly), which will also be done in the study of the CIR model. The main tools used in the study of the impact of correlated noise on the Vasicek model were the gradient boosting method and the LSTM neural network. The results were evaluated using Accuracy, Balanced Accuracy, Precision, and Recall metrics.

D. Cox, J. Ingersoll, and S. Ross, building on the Vasicek model, presented their CIR model (named after the first letters of the authors' surnames), which defines the market price of risk at a given moment in time (Cox et al., 1985). It is also based on the Wiener stochastic process, but now the standard deviation has a multiplier - the square root of the interest rate, which allows for the resolution of the problem of negative interest rates:

$$dr_t = \kappa(\theta - r_t)dt + \sigma\sqrt{(r_t)}\,dW_t, \tag{2}$$

The parameters κ, θ, and σ (which have the same meaning as in the Vasicek model) must be estimated by calibrating the model to market data. κ, θ, and σ have to be estimated by calibrating the model to market data. Usually, the least squares method is used for this purpose (Beshenov, Lapshin, 2019), but in this study, we will use a method based on the maximum likelihood method proposed by G. Orlando, R. M. Mininni, and M. Bufalo (Orlando et al., 2019):

$$\hat{\kappa} = -\ln\left(\frac{(n-1)\sum_{i=2}^{n}\frac{r_i}{r_{i-1}} - \left(\sum_{i=2}^{n}r_i\right)\left(\sum_{i=2}^{n}r_{i-1}^{-1}\right)}{(n-1)^2 - \left(\sum_{i=2}^{n}r_{i-1}\right)\left(\sum_{i=2}^{n}r_{i-1}^{-1}\right)}\right), \tag{3}$$

$$\hat{\theta} = \frac{1}{(n-1)}\sum_{i=2}^{n}r_i + \frac{e^{-\hat{\kappa}}}{(n-1)(1-e^{-\hat{\kappa}})}(r_n - r_1), \tag{4}$$

$$\hat{\sigma}^2 = \frac{\sum_{i=2}^{n} r_{i-1}^{-1} \left(r_i - r_{i-1} e^{-\hat{\kappa}} - \hat{\theta}(1 - e^{-\hat{\kappa}})\right)^2}{\sum_{i=2}^{n} r_{i-1}^{-1} \left(\left(\frac{\hat{\theta}}{2} - r_{i-1}\right) e^{-2\hat{\kappa}} - \left(\hat{\theta} - r_{i-1}\right) e^{-\hat{\kappa}} + \frac{\hat{\theta}}{2}\right)/\hat{\kappa}}. \tag{5}$$

For parameter calibration, monthly U.S. interest rate data from 1995 to 2022 (Federal Funds Effective Rate) were used. The U.S. interest rate data is widely available, reliable, and has extensive historical depth, making it a rich source for analysis. The U.S. economy is one of the largest and most influential globally, influencing interest rates worldwide and making the data particularly representative of financial market dynamics. Additionally, the U.S. interest rate often exhibits distinct behavioral patterns and regime shifts that can be effectively captured by the CIR model, which assumes that interest rates are driven by a mean-reverting process. By focusing on this dataset, the research can leverage established theories and previously documented economic events (e.g., the 2008 financial crisis) as reference points for change point detection. Therefore, in this study, the initial value r_0 is assumed to be 4.1% - the U.S. interest rate as of December 1, 2022.

The classical method for obtaining numerical solutions to stochastic differential equations is the Euler-Maruyama method, a first-order approximation analogous to the Euler method for stochastic differential equations. By applying this method, we can obtain an approximate solution to the stochastic differential equation of the form:

$$r_{i+1} = r_i + \kappa \left(\theta - r_i\right) \Delta t + \sigma \sqrt{r_i} f\left(t\right), \tag{6}$$

where i is the current time step and Δt is the integration step. The key problem with this method is that as a result of this discrete process, r can take negative values with non-zero probability, which prevents further construction of the time series since a negative value would arise under the square root of the equation (Andersen, Jackel, Kahl, 2010). Various studies have proposed different methods to solve this problem, including replacing the expression under the square root with its modulus or setting it to 0 in the case of a negative value. It is also important to note that first-order approximations are easier to analyze mathematically and simpler to apply but are less accurate than higher-order approximations (Kuznetsova, 2013). For the reasons described above, it was decided to consider higher-order approximations, particularly the Milstein scheme:

$$r_{i+1} = r_i + \kappa \left(\theta - r_i\right) \Delta t + \sigma \sqrt{r_i \Delta t} f\left(t\right) + \frac{1}{4}\sigma^2 \Delta t (f^2\left(t\right) - 1). \tag{7}$$

However, this method can also generate negative values for r, which actually contradicts the nature of the original SDE, which guarantees non-negativity - though this is later accompanied by positive drift $\kappa\theta$. As a result, in this study, we decided to use the so-called implicit Milstein scheme:

$$r_{i+1} = \frac{r_i + \kappa\theta\Delta t + \sigma\sqrt{r_i}f(t)\sqrt{\Delta t} + \frac{1}{4}\sigma^2 \Delta t (f(t)^2 - 1)}{1 + \kappa\Delta t}. \tag{8}$$

This method guarantees positive values provided that $4\kappa\theta > \sigma^2$ (Andersen et al., 2010). The calibrated U.S. interest rate parameters satisfy this inequality, thus in our case, all generated series values will be strictly positive when using this method (Table 1).

Table 1. Parameter values based on the empirical data.

Parameter	Value
κ	0.004669
θ	1.390384
σ	0.018495
$4\kappa\theta$	0.025965
σ^2	0.000342

There are many ways to detect change points in a time series. One of the most effective methods for this task is the PELT (Pruned Exact Linear Time) algorithm, which combines accuracy and computational efficiency. It allows the time series to be divided into several homogeneous segments while minimizing a specified loss function. The algorithm is based on dynamic programming and pruning of irrelevant points, which ensures linear time performance in the average case. Let's assume we have a time series, and we want to find such change points that the series can be divided into a certain number of segments. Each segment is characterized by a specific model, such as a constant mean value - accordingly, the tool allows for the identification of four types of changes: mean shift, standard deviation, linear trend, and count. The goal is to minimize the overall loss function, which includes the sum of model errors for each segment and a penalty for the number of segments. The overall loss function for segmenting the time series is given as:

$$\sum_{i=1}^{m+1} [C(y_{(\tau_{i-1}+1):\tau_i}] + \beta pen(m) \tag{9}$$

where C is the segment loss function, and β is the penalty for adding a new change point (regularization parameter). To identify change points, dynamic programming is used. Let $F(s)$ be the minimum loss function for the data from 1 to s. Then:

$$F(s) = \min_{0 \le \tau < s} [F(\tau) + C(y_{(\tau+1):s} + \beta] \tag{10}$$

Initially, $F(0) = -\beta$ (indicating the absence of a segment). Then, starting from the first element of the time series, we sequentially calculate $F(s)$ for all s from 1 to n. The key idea of PELT is that not all points τ need to be considered when calculating $F(t)$. If for some $t' > t$, there exists a τ such that:

$$F(t) + L(y_{t+1:t'}) + \beta < F(t') \tag{11}$$

then point t can be pruned since it will not be optimal for future computations. This significantly reduces computational costs. In the worst case, the complexity of PELT is $O(n^2)$. However, due to effective pruning of irrelevant points, the complexity in practice is often close to $O(n)$, making the algorithm very fast even for large datasets (Fig. 1).

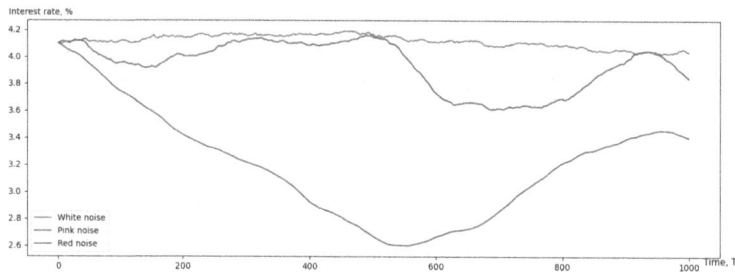

Fig. 1. Example of CIR paths with white, pink, and red noise. (Color figure online)

As previously noted, in practice, models described by stochastic differential equations typically use a Wiener process as the random component, which is a mathematical model of Brownian motion. Normal white noise-a random signal with equal intensity across different frequencies and a constant power spectral density (PSD)-is the derivative of the Wiener process. The Wiener process itself is similar in many ways to white noise: both have constant variance over time and independent increments that follow a normal distribution. The main difference between colored and white noise lies in their spectral characteristics: white noise has a constant PSD across all frequencies, whereas colored noise has a varying PSD depending on the frequency. This makes colored noise preferable in tasks involving the identification and prediction of bifurcation points, as correlations and dependencies can also be observed in real data (Fig. 2).

For analysis, the primary noise colors were selected: pink, red, blue, and violet. Noise color can be characterized by its PSD, given by the equation $1/f^\alpha$ through the parameter α. For pink, red, blue, and violet noises the α parameter is 1, 2, -1, and -2, respectively. Thus, for pink and red noises, the PSD is inversely proportional to frequency, meaning the signal's power density decreases uniformly, while for blue and violet noises, the opposite is true. In addition to colored noises, white noise will also be used to compare the influence of white noise and colored noises on the time series.

3   Results and Discussion

This study investigated the impact of various colored noises—white, pink, red, blue, and violet—on the performance of a change point detection algorithm within time series data. Using the Catboost classifier to discern between normal

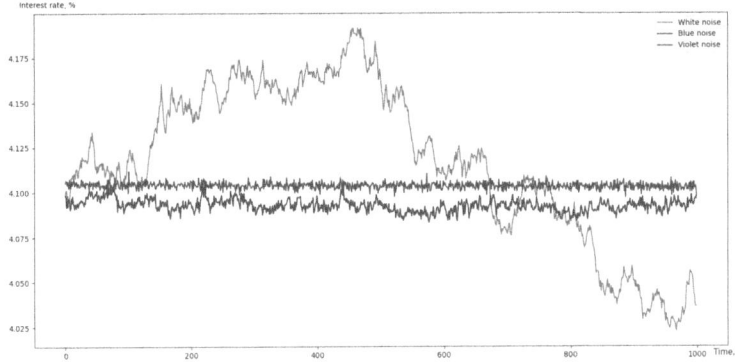

Fig. 2. Example of CIR paths with white, blue, and violet noise. (Color figure online)

points and change points, we evaluated the influence of these noise types through metrics including accuracy, balanced accuracy, precision, and recall. The results offer valuable insights into how different noise characteristics affect algorithmic performance.

The white noise condition, serving as a baseline, yielded suboptimal results across all metrics. These results suggest that white noise, characterized by its equal intensity across all frequencies, significantly challenges the detection algorithm's ability to discern change points, possibly due to its lack of structure and predictability.

In contrast, pink noise, which features a frequency spectrum where power decreases with increasing frequency, demonstrated improved performance. The relatively higher balanced accuracy and recall indicate that pink noise offers a more favorable environment for detecting change points, as its spectral characteristics may introduce a more discernible pattern for the algorithm to leverage. Red noise, or brownian noise, showed comparable performance to pink noise. The results suggest that while red noise's power decreases even more rapidly with frequency compared to pink noise, its impact on the classifier's performance remains similar, indicating a robust capacity of the algorithm to adapt to different noise structures. The blue and violet noise conditions revealed the highest performance metrics, with blue noise (characterized by increased power at higher frequencies) and violet noise (with even more pronounced high-frequency power) achieving accuracy and balanced accuracy of approximately 0.9 for both. Notably, violet noise excelled in all aspects.

In addition to analyzing the impact of different colored noises on change point detection, this study incorporated a parameter η, representing the "change point neighborhood." This parameter adjusts the model to return a detection signal (1) if a change point falls within a specified number of points surrounding the estimated change point. The η values studied were 1, 5, and 10.

The introduction of η did not significantly alter the overall performance metrics across the different noise types. Accuracy, balanced accuracy, recall, and precision remained approximately the same regardless of whether η was set to 1, 5, or 10. This suggests that while the change point neighborhood parameter may provide some flexibility in detecting change points, its impact on the classification performance of both LSTM and Catboost models is minimal in the context of this study (Tables 2, 3).

Table 2. Results of CatBoost model implementation for different parameters η. Here accuracy and balanced accuracy are shown for white and colored noises.

Noise	$\eta = 1$			$\eta = 5$			$\eta = 10$		
Metric	Accuracy								
White	0.58	0.6	0.56	0.6	0.6	0.58	0.56	0.6	0.58
Pink	0.76	0.75	0.8	0.8	0.8	0.76	0.75	0.76	0.8
Red	0.77	0.8	0.81	0.81	0.8	0.77	0.8	0.77	0.81
Blue	0.9	0.9	0.91	0.91	0.91	0.9	0.91	0.9	0.91
Violet	0.89	0.87	0.9	0.9	0.9	0.89	0.87	0.89	0.9
Metric	Balanced accuracy								
White	0.48	0.51	0.5	0.51	0.5	0.48	0.5	0.48	0.51
Pink	0.8	0.77	0.81	0.81	0.81	0.8	0.77	0.8	0.81
Red	0.82	0.82	0.7	0.82	0.82	0.82	0.7	0.82	0.7
Blue	0.83	0.91	0.9	0.91	0.91	0.83	0.9	0.83	0.91
Violet	0.9	0.92	0.91	0.92	0.92	0.9	0.91	0.9	0.92

In parallel with the Catboost model, the same research was conducted using a LSTM model. This analysis introduced a new parameter, P, representing the "length of prediction." The LSTM model predicted whether the point P steps ahead in the time series is a normal point or a change point.

Overall, the performance metrics for the LSTM model remained consistent with those of the Catboost model, with accuracy, balanced accuracy, precision, and recall largely unaffected by the introduction of the P parameter. However, an important observation emerged with the pink and red noise conditions. Specifically, for these types of noise, the LSTM model exhibited a decrease in accuracy and balanced accuracy, both settling around 0.7 (Tables 4 and 5).

The decreased performance could be attributed to the complexities introduced by the longer-term dependencies and the spectral properties of pink and red noise. Despite these challenges, the LSTM model's overall performance across other noise types remained stable, indicating its reliability in change point detection with the exception of specific noise scenarios.

Excellent performance of blue and violet noises is a subject for discussion. Violet and blue noise have a frequency spectrum where power increases with frequency. As a result, violet and blue noise emphasize higher-frequency components of the signal. This can make sudden changes or anomalies more pronounced, as high-frequency changes are more detectable. For change point

Table 3. Results of CatBoost model implementation for different parameters η. Here precision and recall are shown for white and colored noises.

Noise	$\eta = 1$			$\eta = 3$			$\eta = 5$		
Precision									
White	0.48	0.51	0.41	0.48	0.51	0.41	0.48	0.51	0.41
Pink	0.72	0.78	0.76	0.72	0.78	0.76	0.72	0.78	0.76
Red	0.8	0.77	0.74	0.8	0.77	0.74	0.8	0.77	0.74
Blue	0.97	0.87	0.88	0.97	0.87	0.88	0.97	0.87	0.88
Violet	0.95	0.9	0.91	0.95	0.9	0.91	0.95	0.9	0.91
Recall									
White	0.6	0.52	0.47	0.6	0.52	0.47	0.6	0.52	0.47
Pink	0.82	0.81	0.86	0.82	0.81	0.86	0.82	0.81	0.86
Red	0.81	0.8	0.81	0.81	0.8	0.81	0.81	0.8	0.81
Blue	0.84	0.86	0.87	0.84	0.86	0.87	0.84	0.86	0.87
Violet	0.9	0.9	0.87	0.9	0.9	0.87	0.9	0.9	0.87

Table 4. Results of LSTM model implementation for different parameters η and P. Here accuracy and balanced accuracy are shown for white and colored noises.

Noise	$P = 1$			$P = 3$			$P = 5$		
η	1	5	10	1	5	10	1	5	10
Accuracy									
White	0.4	0.49	0.5	0.51	0.49	0.52	0.4	0.41	0.43
Pink	0.71	0.70	0.74	0.74	0.71	0.69	0.71	0.71	0.72
Red	0.76	0.71	0.71	0.67	0.66	0.69	0.73	0.73	0.68
Blue	0.91	0.87	0.98	0.93	0.93	0.9	0.87	0.95	0.91
Violet	0.92	0.91	0.97	0.97	0.95	0.91	0.88	0.99	0.91
Balanced accuracy									
White	0.41	0.43	0.4	0.42	0.43	0.4	0.42	0.39	0.41
Pink	0.7	0.71	0.72	0.71	0.79	0.77	0.63	0.68	0.69
Red	0.77	0.62	0.61	0.66	0.71	0.7	0.69	0.61	0.66
Blue	0.87	0.82	0.83	0.84	0.81	0.84	0.81	0.82	0.84
Violet	0.81	0.91	0.91	0.9	0.88	0.86	0.98	0.92	0.92

detection, this increased emphasis on high frequencies can make it easier for the model to identify shifts or disruptions in the time series. The more pronounced high-frequency components might help the model to recognize patterns indicative of change points more effectively than in the more uniform or irregular noise types like white or pink noise.

Table 5. Results of LSTM model implementation for different parameters η and P. Here precision and recall are shown for white and colored noises.

Noise	P = 1			P = 3			P = 5		
η	1	5	10	1	5	10	1	5	10
Precision									
White	0.41	0.4	0.39	0.39	0.38	0.39	0.4	0.4	0.41
Pink	0.8	0.73	0.68	0.84	0.75	0.67	0.85	0.76	0.66
Red	0.88	0.83	0.79	0.86	0.82	0.77	0.86	0.82	0.77
Blue	0.92	0.89	0.89	0.9	0.87	0.84	0.9	0.88	0.91
Violet	0.91	0.92	0.91	0.98	0.95	0.94	0.88	0.89	0.9
Recall									
White	0.49	0.47	0.44	0.48	0.44	0.43	0.49	0.45	0.43
Pink	0.77	0.84	0.93	0.68	0.71	0.81	0.67	0.61	0.62
Red	0.8	0.87	0.8	0.79	0.78	0.76	0.69	0.77	0.71
Blue	0.9	0.88	0.91	0.92	0.9	0.94	0.88	0.89	0.92
Violet	0.91	0.9	0.92	0.91	0.92	0.96	0.98	0.89	0.9

Relatively low results for pink and red noise (compared to the results of Vasicek model in Zotov et al.) could be provoked by imbalanced dataset. As we can see from the empirical data, change points are rare compared to normal data points, which is in fact an imbalanced dataset that can skew analysis and model training. TimeGAN, a specialized GAN designed for time series data, can offer a promising solution for generating realistic change point features and augmenting the imbalanced dataset (J. Yoon, D. Jarrett, M. van der Schaar). While implementation of TimeGAN is out of scope of this research, it can be seen as an original continuation of studying this topic, as well as whether violet and blue noise are suitable as a replacement of a standard normal noise in a CIR model.

4 Conclusion

This study explored the efficacy of Long Short-Term Memory (LSTM) networks and CatBoost classifiers in detecting change points in time series data subjected to various colored noises: white, pink, red, blue, and violet. The results provide a comprehensive comparison of these methods and insights into why blue and violet noises yield superior performance, while red and pink noises present some challenges.

Both the LSTM and Catboost models demonstrated varied performances across the different types of colored noise. The Catboost classifier, known for its robustness with tabular data, showed a marked performance improvement with structured noises like blue and violet. Specifically, it achieved high accuracy, balanced accuracy, precision, and recall under blue and violet noise conditions, indicating that these types of noise facilitated effective change point detection.

The LSTM model, designed to capture temporal dependencies and patterns, also performed well with blue and violet noise but exhibited some declines in performance when handling pink and red noise. This was particularly evident in accuracy and balanced accuracy, which fell to around 0.7 for pink and red noises, compared to the higher metrics observed with blue and violet noises. The consistent results across different η settings further highlighted that while the change point neighborhood parameter introduced some flexibility, it did not significantly impact the model's performance.

The study underscores the effectiveness of both LSTM and Catboost models in handling different colored noises, with blue and violet noises providing superior conditions for accurate change point detection. While violet noise demonstrates strong potential for enhancing time series analysis, its suitability should be evaluated in the context of specific applications and data characteristics to ensure optimal performance.

These results can provide valuable insights for policymakers, especially in the context of monetary policy and financial stability. As the worth of colored noise regarding time series prediction is justified, by incorporating colored noise into the CIR model, policymakers can better understand how persistent economic disturbances (such as prolonged shifts in market sentiment or structural changes) affect interest rates. This helps in assessing the risks of extended volatility periods and preparing for potential long-term economic shifts. Policymakers can use this knowledge to refine their forecasting models, leading to more informed decisions regarding interest rate adjustments and other monetary policy tools. Further research in this area includes the use of more complex statistical models, such as the Black-Scholes model. Equally important is conducting similar experiments using natural noise, which may more explicitly reflect the instability of financial time series.

References

Beshenov, S., Lapshin, V.: Parametric immunization of interest rate risk via term structure models. HSE Econ. J. **23**(1), 9–31 (2019)

Orlando, G., Mininni, R., Bufalo, M.: Interest rates calibration with a CIR model. J. Risk Financ. (ahead-of-print) (2019). https://doi.org/10.1108/JRF-05-2019-0080

Bibby, B., Jacobsen, M., Sørensen, M.: CHAPTER 4 - Estimating functions for discretely sampled diffusion-type models. In: Handbook of Financial Econometrics: Tools and Techniques, North-Holland, vol. 1 (2010)

Kuznetsova, I.: Numerical solution to stochastic differential equation by euler-maruyama method. Int. Res. J. (11-1(18)), 8–11 (2013)

Andersen, L.B.G., Jäckel, P., Kahl, C.: Simulation of Square-Root Processes. 9780470057568. https://doi.org/10.1002/9780470061602.eqf13009

Orlando, G., Mininni, R.M., Bufalo, M.: A new approach to CIR short-term rates modelling. Contrib. Manag. Sci. **35–43** (2018). https://doi.org/10.1007/978-3-319-95285-72

Orlando, G., Mininni, R., Bufalo, M.: Forecasting interest rates through Vasicek and CIR models: a partitioning approach (2019)

Steffen, D., Andreas, N., Lukasz, S.: An Euler-type method for the strong approximation of the Cox-Ingersoll-Ross process. Proc. R. Soc. A **468**, 1105–1115 (2012). https://doi.org/10.1098/rspa.2011.0505

Zotov, G., Lukianchenko, P.: Change point analysis in Vasicek interest rate model. In: 2023 7th IEEE Congress on Information Science and Technology (CiSt), pp. 19–23 (2023)

Mannella, R.: Integration of stochastic differential equations on a computer. Int. J. Mod. Phys. C - IJMPC **13** (2002). https://doi.org/10.1142/S0129183102004042

Kaulakys, B., Ruseckas, J., Gontis, V., Alaburda, M.: Nonlinear stochastic models of 1/f noise and power-law distributions. Physica A: Stat. Mech. Appl. **365**(1), 217–221 (2006). ISSN 0378-4371, https://doi.org/10.1016/j.physa.2006.01.017

Almeida, C., Lund, B.: Immunization of fixed-income portfolios using an exponential parametric model. Brazil. Rev. Econometrics **34**(2), 155–201 (2014)

Cox, J.C., Ingersoll Jr., J.E., Ross, S.A.: An intertemporal general equilibrium model of asset prices. Econometrica: J. Econometric Soc. 363–384 (1985)

Vasicek, O.: An equilibrium characterization of the term structure. J. Financ. Econ. **5**(2), 177–188 (1977)

Kılıç, D., Uğur, Ö.: Hybrid wavelet-neural network models for time series. Appl. Soft Comput. **144**, 110469 (2023). https://doi.org/10.1016/j.asoc.2023.110469

Borah, J., Chakraborty, T., Mohd Nadzir, M.S., Cayetano, M., Majumdar, S.: WaveCatBoost for Probabilistic Forecasting of Regional Air Quality Data (2024)

A Secure NFC Cardless Cash Withdrawal System

Min-Shiang Hwang[1,2] , Cheng-Ying Lin[3], Chun-Hsien Chang[4],
and Cheng-Ying Yang[5]([✉])

[1] Department of Computer Science and Information Engineering,
Asia University, Taichung, Taiwan, R.O.C.
mshwang@asia.edu.tw
[2] Fintech and Blockchain Research Center, Asia University, Taichung, Taiwan
[3] The Ph.D. Program in Artificial intelligence, Asia University, Taichung, Taiwan
[4] Department of Management Information Systems, National Chung Hsing
University, Taichung, Taiwan, R.O.C.
[5] Department of Computer Science, University of Taipei, Taipei, Taiwan, R.O.C.
cyang@utaipei.edu.tw

Abstract. As IoT technology grows, several different payment methods
are available now. NFC (Near Field Communication) as a short-range
communication technology has been applied in many fields, such as contactless payment and electronic tickets. ATM (Auto Teller Machine) also
benefits from NFC technology as one way of cardless cash withdrawal.
However, there are many security problems, and convenience problems
exist. In security, certain kinds of attacks, such as man-in-the-middle,
replay, and brute-force attacks, must be considered. These attacks may
influence user privacy. Furthermore, considering user experience, existing methods may not make users feel enough convenience when using. To
enhance security and user experience, we proposed a secure NFC cardless
cash withdrawal system using time stamps and nonce to reduce the risk
of attack and design more effortless operations for users to enhance user
experience.

Keywords: NFC · Authentication · Contactless payment · Auto teller
machine

1 Introduction

Due to the widespread use of mobile devices and the improvement of IoT technology, mobile payment is becoming increasingly popular. Short-range communication technology plays an important role in mobile payment [6,10,15,33,35].

The NFC (Near Field Communication) is a technology for short-range communication. NFC uses a 13.56MHz frequency band to make two devices communicate in a range of 20 cm. NFC is developed from RFID technology. Due
to the low computing power of RFID electronic tags, en/decryption operations

K.-W. Huang et al. (Eds.): ICFT 2024, CCIS 2437, pp. 111–121, 2025.
https://doi.org/10.1007/978-981-96-3811-6_11

cannot be performed using traditional cryptographic algorithms, resulting in the leakage of user privacy information [4,13,18,27,32,37].

An NFC device has three work modes: NFC card emulation, NFC reader/writer, and NFC peer-to-peer. NFC card emulation mode enables NFC devices such as smartphones to act like smart cards, allowing users to perform transactions such as payment or ticketing. NFC reader/writer mode enables NFC devices to read information stored on NFC tags embedded in labels or smart posters. Finally, NFC peer-to-peer mode enables two NFC devices to communicate and share information. Because of its short distance of communication and security elements, NFC technology can work well in the payment field with high-security performance [5,22].

ATMs (Automated Teller Machines) also benefit from NFC technology. Traditionally, ATMs require cards to operate. However, nowadays, there are several cardless withdrawal methods. To make cardless withdrawals, one uses NFC-enabled mobile devices, such as a smartphone with a card embedded in it.

However, NFC still suffers from many security problems. For example, security attacks may happen when information exchanges between devices, such as a man-in-middle attack, which harms user privacy and may cause user property loss. Several views need to be considered in an NFC protocol.

Performance: The computation in an NFC protocol for a cardless withdrawal system should not be too complicated because the authentication process in cardless withdrawal should only take a short, and the computing power of mobile devices is limited.

Security: In terms of security, authentication needs to transmit data through the network, such as a man-in-the-middle attack, replay attack, and brute force attack. Otherwise, when users input their private information on their mobile devices, shoulder surfing attacks should be prevented; we also need a way to reduce the risk of missing mobile devices or having them stolen.

In addition to the condition mentioned above, a cardless withdrawal system's user experience is also necessary. An overly annoyed operation may reduce the user's willingness to use it.

A protocol with enough security is necessary to ensure communication safety between smartphones and ATMs. Several existing services and research are aimed at security issues in the NFC field. [7] proposed a protocol that uses two different essential generation methods to ensure security. [11,23] proposed a pseudonym-based NFC protocol. And [1,19–21,24,31] proposed a different scheme to secure passwords used in authentication. [2] proposed an NFC protocol based on a hash function. [26,28] proposed other NFC protocol based on SVM and ECC. In this paper, we propose a lightweight NFC cardless withdrawal system. We use hashed data transfer between the system, time stamp, and nonce generated to prevent the transaction process from being attacked.

The organization of this paper is as follows: Sect. 2 introduces the existing cardless withdrawal method and provides a brief review of different NFC authentication protocols. Section 3 provides our system model. Section 4 is the security and performance analysis of the proposed system. Finally, Sect. 5 presents our conclusion and direction for future work.

2 Related Works

2.1 Existing Cardless Withdrawal Service

Several banks now offer cardless withdrawal services; most of their cardless withdrawal steps are shown in Table 1.

Table 1. Existed cardless withdrawal steps

Device ID	User ID	Hashed ID	Hashed Password	Acc. no.	Acc. balan.
1002053001	16001001	17904686521cda 54adb709670d9e 3542d218f089	121ef27204c69a7 dc2188199569fa7 7c836ccb03	1023402 2031213	23942.00
2202323453	16001002	ce979846ae7990 cbaal17683fle40 5c49cebb5af	20eabe5d6460e21 6796e834f52d615 dc670332fc	1052354 2316212	100233.00

Before using a cardless cash withdrawal service, users must authenticate by going to the bank or using the website or mobile applications offered by the chosen bank.

Whenever users want to use a cardless withdrawal service, they must repeat every step in Table 1. Compared to traditional card cash withdrawal, although it can eliminate the hassle of taking a bank card, these steps may be much more inconvenient than bringing a bank card to withdraw for some users.

2.2 Key Generation Method

Most NFC authentication protocols use keys to ensure the security of communication. Several protocols were based on it. Protocol in [7] using session key generation technology [16] and NTRU public key cryptosystem [12] to secure the payment process. This protocol used session key generation technology to generate and update the session key used in communication and apply the NTRU cryptosystem to encrypt and decrypt payment messages.

In 2016, Feng et al. proposed a lightweight authentication protocol for NFC motion sensor payments [27]. Their solutions are efficient, safe, and easy to implement. The lightweight NFC mobile sensor payment security authentication they proposed matches the HCE trend.

2.3   Pseudonym Method

On the other hand, pseudonym-based NFC protocols are also well-performance NFC authentication [11,23]. In this protocol, the user's identity is represented by a pseudonym, so the attacker cannot get the real user identity. The trusted third party generates a pseudonym.

In 2022, Cao and Liu proposed a lightweight NFC authentication scheme based on an improved hash function to ensure that users' private information will not be leaked [2]. At the same time, their approach is highly efficient and features mutual authentication, forward security, backward security, and security against attacks such as replay, location, and fake. However, Hwang et al. found that once their method becomes out of sync, it must re-establish synchronization data [14]. Therefore, Hwang et al. proposed a lightweight synchronization method.

2.4   Password Secure Method

When we operate our mobile devices to enter PIN codes or passwords, there are risks, such as shoulder surfing or screen record malware. Several types of research are aimed at this problem; the tap-based method [3] records the user's tap melody as a password due to differences between every single person to make sure the security password; the biometric method uses different metrics between each person, such as fingerprint and finger vein as a unique password to enhance security [36]. However, these methods may equip unique accessories on the device, and if it is popularized in the future may be a good choice for authentication.

2.5   Blockchain Method

In 2022, Peng et al. implemented a hierarchical P2P electronic payment system [25] based on the blockchain architecture [8,17,29,30,34,38], called CBP2P (Cooperative Bank Peer-to-Peer), to improve the anonymous crime, extensive data faced by the blockchain financial system Synchronization, and time-consuming transaction flaws. Their approach can achieve a precise electronic payment system and enhance the security and usability of the system through the characteristics of blockchain and cooperation between banks.

3   Secure Cardless Withdrawal System

We will introduce the protocol that we proposed in this section and introduce it in three parts: scheme structure, payment authentication, and verification.

3.1   Scheme Structure

There are three entities in the cardless cash withdrawal system; they are server, ATM, and mobile device. The main functions are described in the following:

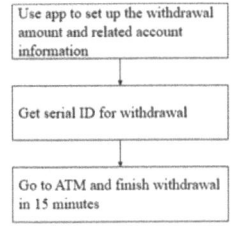

Fig. 1. Sample of data stored in a database.

Server: User information is saved in a database on the server; it contains the device ID, hashed device ID, hashed password, and other related user account information. Figure 1 is the sample that we saved in the database. The hashed device ID is generated in the registration stage. Server and mobile use the same hash function to hash device ID. Hashed ID will help us to enhance efficiency in the authentication process.

ATM: Our system requires an NFC-enabled autoteller machine. In our protocol, the ATM must create a timestamp for authentication between the mobile device and the server.

Mobile device: In our protocol, a mobile device must also be NFC-enabled to communicate with an auto teller machine. Secure elements in mobile devices can store information such as device ID safely.

Figure 2 shows the scheme structure of the proposed secure cardless withdrawal system. We briefly describe it as follows:

1. Registration: Before the secure cardless withdrawal system starts, users (Mobile Device) must register with the server to obtain permission to use the system.
2. Request: The user submits a withdrawal request to the ATM through Mobile Device.
3. Authentication: After the ATM receives the withdrawal request by Mobile Device, it asks the server to verify and confirm the user's identity.
4. Payment: Once the server verifies that the user's identity is correct, the ATM can safely pay money to the user.

3.2   Cardless Withdrawal System

We proposed three stages in the system: Initialization, authentication, and payment. We will introduce it in this section. Three entities are involved in our system: the NFC-enabled mobile device, the NFC-enabled ATM, and the server. Figure 3 shows the proposed cardless withdrawal system scheme. The related symbols used in the proposed scheme are listed in Table 2.

When a user wants to use a cardless withdrawal service, the user has to register by going bank or using the website or another way the bank offered to

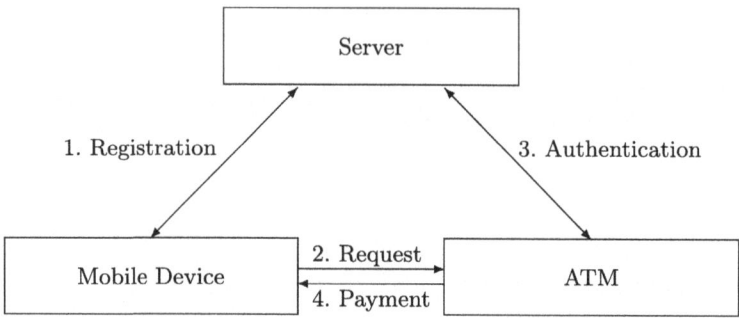

Fig. 2. The scheme structure of the proposed secure cardless withdrawal system.

Table 2. Symbol defifinitions

Symbols	Description
h	hash function;
P	password;
MD	Mobile device;
N	Nonce generated by the mobile device;
T	Time stamp generated by ATM;
R	Result calculated by mobile device and server;
ID	Device ID, which is stored in a secure element;
IDd	Device ID, which is stored in the database by registration.

activate the service. In this stage, necessary user information is stored on the server, such as device id and account information. Notice that the information transmitted in the registration phase should be via the secret channel.

After finishing registration, the user can start using the cardless withdrawal service. First, the user can bring an NFC-enabled mobile device to an ATM and communicate with it through NFC. In this step, the user has to enter the PIN code to activate the service. Next, the ATM generates a time stamp and sends it to the mobile device. Then, the mobile device calculates the time stamp, nonce, and device ID together and sends the result, nonce, timestamp, and hashed ID to the server.

The server will search whether the ID is in the database using the hashed device ID. If the device ID exists, the server will check the result's correctness and send the corresponding information to the mobile device. Finally, mobile devices will check the correctness of the result sent from the server to finish the authentication process. Then the user enters the withdrawal amount on a mobile device, account information, and the amount sent to the ATM via NFC, then sent to the server to finish the withdrawal. After the withdrawal, a message will be sent to the user for confirmation.

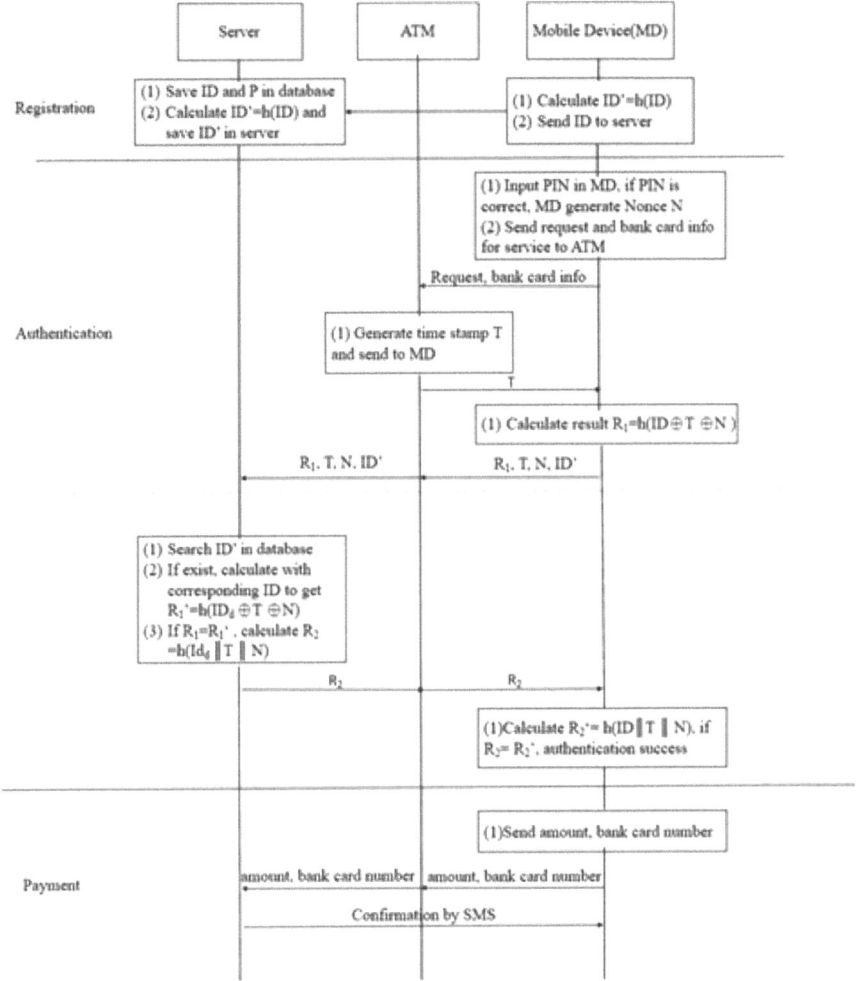

Fig. 3. Cardless withdrawal system scheme.

Initialization stage: The user must finish registration before using the cardless withdrawal service. Sensitive information should be transmitted via secret channels, such as users going to the bank to register or using a website and application offered by the bank and transmitting data via SSL. The user's mobile device ID and related account information will be stored on the server.

Authentication stage: The user brings a mobile device to the ATM to use the cardless withdrawal service; the authentication stage starts after the mobile device communicates with the ATM. Notice that PIN code entry attempts are five times; if the user fails to enter the correct PIN code 5 times, the cardless withdrawal service will be locked temporarily.

Step 1: The user enters the PIN code in the mobile device to activate the authentication process.

Step 2: If activated success, the mobile device generates a nonce N and sends the request and bank card information to the ATM.

Step 3: The ATM generates a time stamp T and sends it to the mobile device.

Step 4: The mobile device calculates device ID, time stamp T, and nonce N together with a hash function and then sends the result and hashed device ID to the server.

Step 5: The server checks whether the ID is in the database by searching the hashed device ID. If the device ID exists in the database, the server calculates the device ID and time stamp T and sends the result to the mobile device.

Step 6: The mobile device checks the result from the server and finishes the authentication stage.

Payment stage: After the authentication stage, users can enter the amount using a mobile device. The related account information is delivered to the ATM via NFC and then passed to the server. After the withdrawal process, a message with transaction details such as account available balance, amount, bank ID, ATM ID, date, and time is sent to the user's device.

4 Security Analysis of Cardless Withdrawal System

High security is necessary as an NFC authentication protocol aimed at cardless withdrawal, which is highly related to personal privacy and property. Therefore, we will analyze our system from the view of security in this section.

4.1 Security Analysis

Brute Force Attack. Due to the time stamp generated by ATM and nonce generated by mobile device changes every time and the ID are stored in a secure element safely, our system can enhance resistance to brute force attacks. In addition, the attempt limit can also vastly reduce the risk of brute force attacks regarding PIN codes.

Man-in-the-Middle Attack. The attacker can't access user privacy identity information because the device ID will not be transferred in plain text. However, hashed information and the result may be captured; an attacker cannot use its login system.

Replay Attack. By using nonce generated by the mobile device and time stamp, our proposed system can resist replay attacks. Although attackers may be able to capture the data transmitted in the system, replaying these data cannot pass system verification successfully.

Card Cloning Attack. Some attackers tried to record the card information and create a copy to use. In our proposed system, only using the copy of the card embedded in a mobile device cannot pass the authentication stage. Several necessary pieces of information for authentication are stored in the secure element.

Shoulder Surfing Attack. The attacker tries to observe the user enter the password and remember it for later use. In our proposed system, only remembering the password can't pass authentication. For more security, we use random keyboards in our system to improve resistance to shoulder surfing attacks.

Theft Mobile Device. Our system needs to enter a PIN code before activating the system. If an attacker attempts to guess the PIN code, the attempt limit will prevent these guesses. So that only getting a mobile device cannot access the system successfully.

4.2   Performance Analysis

Besides security, performance is also an essential point in the system. Each entity's computing power is limited, and the cash withdrawal system needs an instant reaction. Compared to existing services offered by banks (we take the service offered by TaiShin Bank as an example), users need to operate their bank's application and go to the ATM to finish the cardless withdrawal process in 15 minutes. On the other hand, our proposed scheme only requires bringing a mobile device to the ATM to finish authentication and get cash instantly.

Compare the computation complication with the scheme used in payment authentication we mentioned in the previous part. However, as shown in Table 3, our scheme can consume less computing power.

Table 3. Comparison with other NFC protocol

Schemes	Encryption	Decryption	Hash Function
Fan et al.	3	3	1
The Proposed Method	-	-	2

5   Conclusions

Firstly, we proposed a secure cardless withdrawal system. The system consists of three entities: server, ATM, and mobile device, and it is designed from three points of view: registration stage, authentication stage, and payment stage. Our system cancels the pre-activation step on most existing services, which may cause users to consider that withdrawing with a card is more convenient than a cardless withdrawal. On the other hand, security is enough between entities to resist specific attacks. Our proposed system is aimed at a cash withdrawal system now; in future work, we will try to apply it to other NFC payment situations and combine some new technology, such as blockchain.

References

1. Cao, F.M., He, X.P.: Lightweight RFID bidirectional authentication protocol based on improved hash function. Int. J. Netw. Secur. **26**(1), 98–105 (2024)
2. Cao, F.M., Liu, D.W.: A lightweight NFC authentication algorithm based on modified hash function. Int. J. Netw. Secur. **24**(3), 436–443 (2022)
3. Chen, Y., Sun, J., Zhang, R., Zhang, Y.: Your song your way: rhythm-based two-factor authentication for multi-touch mobile devices. In: IEEE Conference on Computer Communications (INFOCOM'15), pp. 2686–2694. IEEE (2015)
4. Chen, Y.C., Wang, W.L., Hwang, M.S.: RFID authentication protocol for anti-counterfeiting and privacy protection. In: The 9th International Conference on Advanced Communication Technology, pp. 255–259 (2007)
5. Chi, Y.L., Chen, C.H., Lin, I.C., Hwang, M.S.: The secure transaction protocol in NFC card emulation mode. Int. J. Netw. Secur. **17**(4), 431–438 (2015)
6. Dong, X., Li, J.: A blockchain technology: analysis of a secure payment mode for enterprise e-commerce import and export trade. Int. J. Netw. Secur. **26**(5), 861–866 (2024)
7. Fan, K., et al.: NFC secure payment and verification scheme with CS E-ticket. Secur. Commun. Netw. **2017** (1), Article ID 4796373 (2016)
8. Feng, D., Li, H.: Research on education information security sharing based on blockchain and Bayesian network. Int. J. Netw. Secur. **26**(1), 25–32 (2024)
9. Feng, T.H., Hwang, M.S., Syu, L.W.: An authentication protocol for lightweight NFC mobile sensors payment. Informatica **27**(4), 723–732 (2016)
10. Gao, W.J., Yue, Y.Q., Wang, X.Q.: Design and implementation of a central node-controlled off-chain payment channel rebalancing scheme. Int. J. Netw. Secur. **26**(2), 257–269 (2024)
11. He, D.B., Kumar, N., Lee, J.H.: Secure pseudonym-based near field communication protocol for the consumer internet of things. IEEE Trans. Consum. Electron. **61**(1), 56–62 (2015)
12. Hoffstein, J., Pipher, J., Silverman, J.H.: NTRU: a ring-based public key cryptosystem. In: Buhler, J.P. (eds.) ANTS 1998. LNCS, vol. 1423. Springer, Heidelberg (1998). https://doi.org/10.1007/BFb0054868
13. Hsieh, Y.Y., Chang, L.H., Liao, A.Y.H., Yang, C.Y., Hwang, M.S.: The system adoption evaluation of RFID safety management system on campus. Int. J. Netw. Secur. **24**(1), 176–180 (2022)
14. Hwang, M.S., Li, H.W., Yang, C.Y.: An improvement of a lightweight NFC authentication algorithm based on modified hash function. In: The 14th International Conference on ICT Convergence, Lotte Hotel, Jeju Island, Korea (2023)
15. Kang, B., Du, J., Han, Y., Qian, K.: A lightweight anonymous mobile payment scheme for digital commodity in cloud computing service. Int. J. Netw. Secur. **22**(6), 945–953 (2020)
16. Kungpisdan, S., Metheekul, S.: A secure offline key generation with protection against key compromise. https://api.semanticscholar.org/CorpusID:18344266. Accessed 9 June 2024
17. Li, D.: Research on blockchain technology for security protection of network information data in the legal system background. Int. J. Netw. Secur. **26**(2), 173–179 (2024)
18. Li, Y.Z., Zuo, W.T., Liu, D.W.: Tag group coexistence protocol for verifiable RFID system. Int. J. Netw. Secur. **24**(6), 1056–1063 (2022)

19. Li, Z.H.: Mobile RFID authentication protocol based on parity check patching operation. Int. J. Netw. Secur. **25**(6), 920–927 (2023)
20. Liu, L., Jia, Y.: Analysis of one lightweight authentication and matrix-based key agreement scheme for healthcare in fog computing. Int. J. Netw. Secur. **26**(1), 138–141 (2024)
21. Liu, W.R., Ji, Z.Y., Chu, C.C.: An improved secure RFID authentication protocol using elliptic curve cryptography. Int. J. Netw. Secur. **26**(1), 106–115 (2024)
22. Ma, Y.: NFC communications-based mutual authentication scheme for the internet of things. Int. J. Netw. Secur. **19**(4), 631–638 (2017)
23. Odelu, V., Das, A.K., Goswami, A.: SEAP: secure and efficient authentication protocol for NFC applications using pseudonyms. IEEE Trans. Consum. Electron. **62**(1), 30–38 (2016)
24. Pan, T., Zuo, K.Z., Wang, T.C., Deng, C.H.: A lightweight authentication protocol for mobile RFID. Int. J. Netw. Secur. **26**(2), 190–199 (2024)
25. Peng, Z.P., Chiang, M.L., Lin, I.C., Yang, C.C., Hwang, M.S.: CBP2P: cooperative electronic bank payment systems based on blockchain technology. J. Internet Technol. **23**(4), 683–692 (2022)
26. Wang, Z., Wang, W., Zhang, J., Yang, T.: NFC-defender: SVM-based attack detection on NFC-enabled mobile device. Int. J. Netw. Secur. **23**(3), 379–385 (2021)
27. Wei, C.H., Hwang, M.S., Chin, A.Y.H.: A mutual authentication protocol for RFID. IEEE IT Prof. **13**(2), 20–24 (2011)
28. Wei, Y., Chen, J.: Tripartite Authentication Protocol RFID/NFC based on ECC. Int. J. Netw. Secur. **22**(4), 664–671 (2020)
29. Wu, C.C., Chang, C.S., Lin, I.C., Hwang, M.S.: Research on blockchain secret key sharing and its digital asset applications. Int. J. Netw. Secur. **26**(1), 160–166 (2024)
30. Wu, S., Liu, Q., Li, J.: Research on the detection of illegal transactions in currency transactions based on blockchain technology. Int. J. Netw. Secur. **26**(1), 19–24 (2024)
31. Xu, S.H.: Three-party authentication protocol based on Riro for mobile RFID system. Int. J. Netw. Secur. **26**(1), 1–9 (2024)
32. Xu, S.H., Liu, D.W., Zuo, W.T.: A mobile RFID authentication protocol based on self-assembling cross-bit algorithm. Int. J. Netw. Secur. **25**(1), 1–9 (2023)
33. Yan, F., Xing, L., Zhang, Z.: An improved certificateless signature scheme for IoT-based mobile payment. Int. J. Netw. Secur. **23**(5), 904–913 (2021)
34. Yan, Y., Liu, Q., Li, J.: Blockchain collaborative coin mixing scheme based on hierarchical mechanism. Int. J. Netw. Secur. **26**(3), 442–453 (2024)
35. Yang, T.: Mobile payment security in the context of big data: certificateless public key cryptography. Int. J. Netw. Secur. **22**(4), 621–626 (2020)
36. Yang, W.C., Hu, J.K., Wang, S., Wu, Q.H.: Biometrics based privacy-preserving authentication and mobile template protection. Wirel. Commun. Mob. Comput. **2018**(1), Article ID 7107295 (2018)
37. Zhan, S.H., Yu, C.Q.: Mobile RFID authentication protocol based on permutation cross synthesis for anti counterfeit attack. Int. J. Netw. Secur. **24**(2), 305–313 (2022)
38. Zhao, J., Zhu, J.: Data privacy protection based on unsupervised learning and blockchain technology. Int. J. Netw. Secur. **26**(2), 312–320 (2024)

FinTech Digital Transformation: Generative AI, Humanoid Robots, Metaverse, Human-AI Collaboration, and Industry 5.0

Yuxin Liu[1,2] , Runyu Wang[1,3] , and Keng Siau[4(✉)]

[1] City University of Hong Kong, Hong Kong, Hong Kong SAR
[2] Tianjin University, Tianjin, China
[3] Harbin Institute of Technology, Harbin, China
[4] Singapore Management University, Singapore, Singapore
klsiau@smu.edu.sg

Abstract. This paper discusses the transformative impact of emerging digital technologies on the digital transformation of the financial industry, focusing on integrating Generative AI (GenAI), humanoid robots, and the Metaverse within the framework of Industry 5.0. Industry 5.0 emphasizes a human-centric approach to technology, where human-AI collaboration plays a central role in reshaping financial services. By reviewing both academic research and practical applications, the paper highlights the current advancements in FinTech, particularly in AI technologies and the Metaverse, and their future potential, demonstrating how these innovations are driving growth, efficiency, and resilience in the financial sector. Further, the paper proposes a three-phase research framework aimed at addressing the technological, legal, and ethical challenges associated with digital transformation. This framework focuses on key issues such as stability, security, data privacy, and equitable access to digital financial services. Ultimately, the paper argues that the integration of advanced digital technologies with human expertise and sustainable practices has the potential to revolutionize the global financial ecosystem, fostering a more adaptive, secure, and equitable industry in the future.

Keywords: FinTech · Artificial Intelligence · Metaverse · Human-AI Collaboration · Industry 5.0

1 Introduction

1.1 Digital Transformation in the Financial Industry

Technological advancements and innovations have elevated the financial technology (FinTech) sector from a peripheral role to a pivotal force in financial services in the past decades (Siau et al., 2018). The development of blockchain, digital banking, and AI technologies like robo-advisors, InsurTech, and PropTech are revolutionizing the financial landscape. The global FinTech market, valued at USD 294.74 billion in 2023, is expected to grow to USD 340.10 billion in 2024 and is projected to reach USD 1,152.06 billion by 2032 (Fortune Business Insight, 2023).

© The Author(s), under exclusive license to Springer Nature Singapore Pte Ltd. 2025
K.-W. Huang et al. (Eds.): ICFT 2024, CCIS 2437, pp. 122–136, 2025.
https://doi.org/10.1007/978-981-96-3811-6_12

FinTech digital transformation, or digital transformation in the financial industry, refers to the strategic integration of advanced digital technologies into core financial services. This transformation aims to improve operational efficiency, enhance customer experiences, and drive innovation, fundamentally reshaping how financial institutions operate (Riasanow et al., 2018). Moreover, digital transformation also catalyzes a cultural evolution in the financial industry, necessitating a paradigm shift towards a more agile and innovative mindset (Fitzgerald et al., 2014).

With rapid advancements in cutting-edge technologies such as artificial intelligence (AI) and Metaverse, financial institutions are increasingly recognizing the need to adapt their business models to remain competitive in a dynamic market environment (Bharad-waj et al., 2013; Fitzgerald et al., 2014). Therefore, financial institutions are actively developing comprehensive digital strategies to navigate the complexities of DT and guide the entire organization through the transformation (Gobble, 2018). Governments around the world also promote regulatory frameworks and incentives to support Fin-Tech's digital transformation, fostering an environment conducive to innovation and growth. However, digital transformation journeys still pose significant challenges for the financial industry, such as aligning technology innovations with business strategies, ensuring employee engagement and skill development, and addressing evolving customer expectations and user adoption hurdles (Fitzgerald et al., 2014; Nambisan et al., 2019; Brunetti et al., 2020).

1.2 Cutting-Edge Technologies in FinTech Digital Transformation

The rise of FinTech has transformed traditional financial services by enabling seamless transactions, personalized financial advice, and enhanced risk management (Arner et al., 2015; Chen, 2024). AI, in particular, plays a pivotal role in optimizing financial decisions, improving product recommendations, and automating complex financial operations.

Among the various advancements in AI, Generative AI (GenAI) represents one of the most advanced developments with great potential to revolutionize the delivery of financial services. By integrating GenAI technologies, financial institutions can automate more complex tasks, such as data analysis and report generation, allowing human experts to focus on more strategic decision-making. Generative AI can also create personalized financial advice by analyzing individual user data and preferences, thus tailoring recommendations that meet the unique needs of each client. This ability to customize interactions not only improves user experience but also enhances the overall effectiveness of financial services.

Moreover, humanoid robots are emerging as significant players within the financial industry, offering human-like interactions in customer service and advisory roles (Cao, 2024). Humanoid robots are AI-driven robots resembling the human body in shape and imbued with emotional intelligence, empathy, and social skills (Obaigbena et al., 2024). These robots can engage customers in a conversational manner, providing instant responses to inquiries and facilitating seamless transactions. Users can be easier to navigate complex financial products and services with humanoid robots. As users become more familiar with these humanoid interfaces, their potential for improved customer engagement and loyalty will increase.

The emergence of Metaverse also opens new possibilities for both consumers and financial institutions. The Metaverse, sometimes known as the next Internet, is an emerging virtual environment where digital assets and digital identities converge, creating immersive digital experiences (Wang et al., 2024; Birch & Richardson, 2023). This dynamic concept is fueled by the integration of various technologies, including virtual reality (VR), augmented reality (AR), mixed reality (MR), blockchain, and artificial intelligence (AI) (Yang et al., 2022; Duan et al., 2021). Blockchain-based transactions and decentralized finance (DeFi) are emerging as key components within the Metaverse, enabling activities such as virtual trading, digital collectibles, and decentralized financial ecosystems, offering users greater autonomy and transparency (Huang et al., 2023). Recent research has emphasized how AI drives more immersive interactions and enhances user experiences by personalizing content and optimizing virtual ecosystems (Fadhel et al., 2024). In this virtual realm, FinTech solutions are embedded within digital ecosystems, offering benefits such as frictionless financial transactions, personalized user experiences, and innovative financial services. As the Metaverse continues to evolve, FinTech's role within it promises to revolutionize the global financial landscape, enabling decentralized, efficient, and interactive financial activities. The integration of AI-driven FinTech solutions within virtual ecosystems signals the advent of a new era, offering opportunities to reshape finance in innovative and sustainable ways.

1.3 Human-AI Collaboration in Industry 5.0

The concept of Industry 5.0 emerges as a response to the limitations of Industry 4.0, which focuses primarily on automation and efficiency but has ignored many sustainability concerns (Ghobakhloo et al., 2023). Industry 5.0 shifts the focus towards a more human-centric perspective, advocating for enhanced collaboration between humans and machines (Adel, 2022; Soomro et al., 2022). This new paradigm aims to harness the capabilities of advanced technologies, such as AI and robotics, not just for economic gains but to prioritize sustainability, ethical considerations, and the enhancement of human well-being (Adel, 2022; Raja Santhi & Muthuswamy, 2023). Industry 5.0 aims to create value not only for organizations but also for society at large, aligning business objectives with broader social and environmental goals.

Human-AI collaboration plays a central role in this evolution, enabling the integration of human creativity, critical thinking, and emotional intelligence with machine efficiency, thereby fostering innovation and improving decision-making processes and outcomes (Shneiderman, 2020; Raja Santhi & Muthuswamy, 2023). As digital technologies continue to evolve at a rapid pace, financial institutions must not only leverage these innovations to boost operational efficiencies but also develop sustainable human-AI collaboration strategies that ensure technology empowers—rather than replaces—human expertise. A key priority for the financial industry, in alignment with Industry 5.0, is to create a more engaging, transparent, and human-centric financial environment.

To summarize, cutting-edge technologies offer significant opportunities but also pose numerous challenges for the financial industry. Governments, financial institutions, technology companies, and other stakeholders are striving to harness these opportunities while addressing the challenges to foster the development of the financial sector within the framework of Industry 5.0. This study aims to provide valuable insights into this process. Following a literature review and an analysis of current advancements in digital technologies, we propose research opportunities and future directions from the perspectives of law, technology, and ethics.

2   Literature Review

2.1   FinTech Development: Generative AI and Humanoid Robots

FinTech, propelled by advancements in AI and digital technologies, has been a central focus of recent research. Previous studies have categorized the FinTech ecosystem into eight key segments: payments and money transfers, capital markets innovations (e.g., algorithmic trading, high-frequency traders, and market analytics), digital banking, digital wealth managers (e.g., robo-advisors), FinTech lending (e.g., marketplace lenders), equity crowdfunding, InsureTech (i.e., innovations in the insurance industry), and PropTech (i.e., innovations in the property and real estate industry) (Imerman & Fabozzi, 2020). Research has investigated each segment, emphasizing customer adoption, usage behavior, and post-use attitudes, providing valuable insights into their development.

The impact of AI on financial services is highlighted through key applications such as algorithmic trading, volatility forecasting, portfolio management, and investor sentiment analysis (Bahoo et al., 2024). Their comprehensive analysis underscores how AI-driven systems are reshaping financial markets through advanced predictive and classification tools, enhancing firms' ability to manage risk and improve performance. Further discussion on the regulatory and ethical challenges that accompany AI adoption in finance emphasizes the need for governance frameworks to address issues like data privacy, security, and accountability in FinTech applications such as fraud prevention, credit risk assessment, and investment management (Ridzuan et al., 2024).

GenAI and AI-enhanced humanoid robots represent the latest AI advancements in the FinTech sector. GenAI has shown immense potential in financial analytics, providing innovative solutions for various applications, including fraud detection, risk management, predictive analytics, and behavioral analysis (Saivasan & Lokhande, 2023). By embracing generative AI, financial institutions can further automate and optimize delivery processes of financial services, enhance user experiences, and drive innovation in financial products and services.

Robo-advisors have the capability to support their users in financial decision-making, like risk measurement, portfolio selection, and rebalancing (Jung et al., 2019). In the context of humanoid robots, research has identified key factors driving the adoption of robo-advisors, including attitudes, subjective norms, and sociodemographic variables (Belanche et al., 2019). Further expanding this understanding, research demonstrates that consumers exhibit positive attitudes toward robo-advisor services, with perceived ease of use, usefulness, and convenience playing critical roles (Sabir et al., 2023). These studies underscore the transformative role of humanoid robots in enhancing financial services and ensuring operational efficiency.

2.2   Metaverse

The Metaverse is a post-reality universe that merges physical reality with digital virtuality, creating a seamless, persistent, and multiuser environment. It is driven by the convergence of technologies such as virtual reality (VR) and augmented reality (AR), allowing users to interact with digital objects and environments in real time (Lee et al., 2021; Mystakidis, 2022). A three-layer architecture—comprising infrastructure, interaction, and ecosystem—frames the development of the Metaverse (Duan et al., 2021). Moreover, Huang et al. (2023) provided a systematic overview of the economic systems within the Metaverse, discussing the frameworks that connect digital creation, digital assets, and virtual trading markets.

Scholars have explored the promise of the Metaverse across various domains, such as gaming and entertainment (Swami, 2024), education and training (Kye et al., 2021; Jovanović & Milosavljević, 2022), work and collaboration (Hennig-Thurau et al., 2023; Schiller et al., 2024), retail and e-commerce (Popescu et al., 2022), healthcare and therapy (Ford et al., 2023), real estate (Ante et al., 2023), finance and investment (Şanlısoy & Çiloğlu, 2023), and tourism and exploration (Buhalis et al., 2023). These advancements collectively demonstrate the potential of the Metaverse to revolutionize virtual interactions while also highlighting the significant technological, legal, and ethical challenges that must be addressed to ensure scalability and accessibility for a global audience.

2.3   FinTech in Metaverse

The Metaverse is shaping the future of finance by creating new opportunities for innovation, efficiency, and engagement within virtual environments. This convergence of financial activities and Metaverse is not only transforming customer experiences but also driving the development of decentralized financial systems and digital assets. Current academic work highlights the growing significance of Metaverse within the financial industry. For instance, Ooi et al. (2023) examined how the Metaverse could reshape various sectors of banking, including corporate banking, retail banking, and the roles of employees, as well as its potential implications for public policy, revealing the transformative potential of virtual financial ecosystems. They argue that the integration of immersive technologies within banking can facilitate more meaningful interactions between institutions and their clients, leading to a more engaged and informed customer base (Stephanidis et al., 2019).

With lessons from information technology's disruptive and transformative potential, such as the Internet, banking and financial services are eyeing the Metaverse and exploring ways to gain competitive advantages in this new business environment (Siau, 2003). The next generation of the Internet of Things (IoT) ecosystem, comprising technologies such as AI, 5G/6G, the Metaverse, and digital twins, is expected to become the foundation of global finance and business by 2030, in alignment with the UN's Sustainable Development Goals (SDGs) (Bhat et al., 2023). Further investigation into decentralized finance (DeFi) technologies reveals how they facilitate seamless financial operations in virtual environments, particularly when empowered by AI (Far et al., 2023).

Moreover, AI- and robot-powered financial services are becoming increasingly prevalent in the Metaverse. These services facilitate customer interactions, handle transactions, and provide financial advice in real time. However, with this rapid transformation, ethical considerations around data privacy and algorithmic bias remain critical concerns, as highlighted by the recent focus on data protection regulations (Wirtz & Pitardi, 2023; Sautunnida et al., 2023). Institutions are now tasked with not only leveraging these technologies but also ensuring that their implementation remains transparent, equitable, and compliant with emerging regulations.

3   Current Development of FinTech and Metaverse

3.1   Current Development of FinTech: Generative AI and Humanoid Robots

The growing integration of advanced AI technologies, including GenAI and humanoid robots, is playing a significant role in the ongoing digital transformation of the financial industry. The GenAI market in finance is expected to experience substantial growth, with projections showing an increase from $1.09 billion in 2023 to more than $12 billion by 2033[1]. GenAI is revolutionizing financial processes by automating data analysis, providing real-time insights, and enabling personalized financial products through natural language generation (NLG) and machine learning. For instance, GenAI is not just automating processes but evolving how financial institutions extract meaningful insights from large datasets, such as earnings calls and financial news, by capturing subtle nuances in management's communication and market sentiment, as noted by Dennis Walsh, global co-head, Quantitative Investment Strategies, Goldman Sachs Asset Management[2]. These advanced AI systems enhance critical tasks such as risk management, fraud detection, and customer service.

Humanoid robots also gain increasing attention in the financial industry, as they are capable of providing more intuitive and human-like interactions in customer service settings. The global market size is predicted to achieve $38 billion by 2035[3]. In the financial industry, for example, HSBC Bank USA, in partnership with SoftBank Robotics America, has deployed Pepper, a social humanoid robot, in multiple branches across the U.S., including Miami and Seattle. This rollout marks a milestone in retail banking, where Pepper improves customer engagement by offering personalized assistance, educating customers on financial products, and facilitating self-service options. The robot uses AI to ask relevant questions, determine customer needs, and enhance the overall banking experience with intuitive, human-like interactions[4]. By incorporating humanoid robots, financial institutions can improve client engagement by offering

---

[1] https://www.statista.com/statistics/1449285/global-generative-ai-in-financial-services-market-size/.

[2] https://www.goldmansachs.com/insights/articles/how-generative-ai-tools-are-changing-systematic-investing.

[3] https://www.goldmansachs.com/insights/articles/the-global-market-for-robots-could-reach-38-billion-by-2035.

[4] https://fintechmagazine.com/banking/hsbc-rolls-out-softbanks-robot-pepper-achieve-branch-future.

personalized support and advice, mimicking human behaviors to create a more natural interaction environment.

3.2 Current Development of Metaverse

The Metaverse is evolving and maturing rapidly. Countries worldwide are recognizing the enormous potential of the Metaverse and are making substantial efforts to drive its development (Wang et al., 2024). Economic investments in the Metaverse are increasingly visible on the international stage. Leading technology companies, like Meta and Microsoft, are making heavy investments in Metaverse technologies and actively promoting collaborations[5]. Seoul Metropolitan Government (SMG) has invested over 3.8 billion dollars to build Meta Seoul as part of its "Seoul Vision 2030" plan[6].

Technological advancements in the Metaverse are progressing across multiple dimensions. Innovations in virtual reality (VR) and augmented reality (AR) hardware (e.g., Meta Quest 3) have significantly enhanced the accessibility and realism of the Metaverse, enabling users to engage seamlessly with immersive 3D environments. Technologies supporting digital humans and avatars in the Metaverse (e.g., motion capture and 3D modeling) have provided users with enhanced immersion and greater diverse experiences within virtual spaces[7]. In addition, the extensive potential of digital environments in fostering social and collaborative interactions is increasingly being explored. Platforms such as Decentraland and Horizon Worlds have emerged as central hubs for a variety of virtual activities, extending beyond mere entertainment to include meetings, marketing, and educational activities (Ng, 2022).

The government and relevant institutes also actively promote political initiatives to guarantee the healthy and smooth development of Metaverse. For instance, an initiative focused on virtual worlds and the Metaverse was proposed under the framework of "A Europe Fit for the Digital Age", aiming to advance Europe's goals of achieving digital sovereignty by 2030[8].

3.3 Current Development of FinTech in Metaverse

Blending the advanced AI algorithms with the data-rich environment of the Metaverse has great potential to enhance financial activities. In addition, the immersive environment and AI-driven avatars in the Metaverse facilitate virtual 'face-to-face' interactions in financial scenarios. Financial institutions worldwide are closely following the wave of Metaverse and AI to seize novel opportunities of FinTech and reshape the financial landscape. For instance, South Africa's Nedbank, one of Africa's largest financial services groups, has also established a village within Africarare, South Africa's first Metaverse,

---

[5] https://mpost.io/wp-content/uploads/Deloitte.pdf.

[6] https://www.worldexcellence.com/seoul-is-the-first-city-to-enter-the-metaverse-all-about-the-seoul-vision-2030-plan/.

[7] https://www.synthesia.io/post/digital-humans.

[8] https://www.globalpolicywatch.com/2023/04/regulating-the-metaverse-in-europe/.

which offers users opportunities for various economic transactions, including land purchases and NFT collections[9]. Standard Chartered Bank (SCB) has also purchased a
virtual land in The Sandbox, which is a blockchain-based virtual world with many real-
world brands engaged. Governments are also playing an active role in shaping the future
of such integration (Wang & Siau, 2019). For instance, the Hong Kong government
has issued the Policy Statement on the Development of Virtual Assets in Hong Kong[10],
which emphasizes the government's policy supports and guidelines on the development
of the Metaverse and the virtual asset industry.

3.4 FinTech Digital Transformation in the Context of Industry 5.0

The framework of Industry 5.0 is playing a crucial role in facilitating the FinTech digital transformation, guiding the intersection of Generative AI, humanoid robots, and
the Metaverse within the financial industry. Industry 5.0 emphasizes the collaboration
between humans and advanced technologies, ensuring that technological advancements
enhance operational efficiency and create a more personalized, human-centered user
experience. Thus, in the context of Industry 5.0, the FinTech digital transformation
is reshaping the financial industry by creating a more human-centric, sustainable, and
innovative environment.

Human-AI collaboration allows financial services to be more responsive and personalized while also addressing broader social and environmental goals by fostering more
ethical and sustainable financial practices. A prominent practice of human-AI collaboration is the use of hybrid models in financial services. For example, platforms like Better-
ment and Vanguard utilize robo-advisors to manage customer portfolios based on algorithms that allocate assets according to risk profiles and goals. However, these services
also provide access to human financial advisors, allowing for a hybrid model where customers benefit from AI-driven efficiency while still receiving personalized advice from
humans when needed. This blending of automated and human advice demonstrates how
human-AI collaboration enhances service quality in investment advisory. The human-AI
collaboration fosters creativity and innovation, particularly in complex financial services,
such as wealth management and investment advisory, where human intuition combined
with AI-driven data analytics creates more value for customers. Financial institutions are
rapidly adopting these technologies to increase operational efficiency, drive innovation,
and enhance customer experiences (Raja Santhi & Muthuswamy, 2023). By integrating
AI, humanoid robots, and Metaverse technologies, financial institutions are not only
advancing their digital transformation but also contributing to the development of an
interconnected ecosystem where digital and physical worlds merge, providing seamless
and innovative financial solutions for users in an increasingly complex global market.

---

[9] https://www.wipro.com/banking/the-next-phase-of-metaverse-banking-from-conception-to-
reality/

[10] https://gia.info.gov.hk/general/202210/31/P2022103000454_404805_1_1667173469522.pdf.

4   Research Opportunities and Directions

Cutting-edge technologies present numerous opportunities for the evolution of the financial industry. By seizing these opportunities, financial institutions can position themselves advantageously within the competitive landscape. Furthermore, existing research underscores the importance of fostering sustainable growth and resilience in line with Industry 5.0 principles. Organizations embracing Industry 5.0 principles can achieve enhanced adaptability and responsiveness to market changes, ultimately leading to improved customer satisfaction and loyalty (Raja Santhi & Muthuswamy, 2023). Furthermore, the integration of advanced technologies within a human-centric framework also cultivates innovation ecosystems, promoting collaboration among diverse stakeholders, including businesses, governments, and communities. As such, Industry 5.0 presents a compelling opportunity for financial institutions to refine their operational strategies and contribute positively to societal well-being in an increasingly complex and interconnected world.

However, despite significant advancements in FinTech and Metaverse, the journey toward digital transformation in the financial industry is far from complete. The heightened human-centric standards of Industry 5.0 present considerable challenges for integrating advanced technologies into established practices. A gap persists between the visionary potential of these innovations and the practical realities, spanning technical, legal, and ethical dimensions. Addressing these challenges requires a comprehensive approach to digital transformation that not only promotes technological innovations but also aligns with the increased demands of human-AI collaboration and the evolving expectations of the broader societal context (Nah et al., 2023). To bridge this gap, stakeholders must address key issues to facilitate effective and secure integration. This section discusses challenges to FinTech digital transformation to Industry 5.0 based on three perspectives: technological, legal, and ethical considerations, and proposes a three-phase framework for research opportunities and directions.

4.1   Technological Perspective

The immaturity of current AI technologies remains a significant challenge that hinders the progress of digital transformation in the financial industry. While GenAI has demonstrated impressive capabilities in engaging human-like conversations and creating various forms of content such as text, images, videos, and code, widespread concerns over the accuracy of its responses persist, highlighting the need for further refinement and development of GenAI technologies. In addition, the interface design of AI technologies, including GenAI and Humanoid Robots, plays a crucial role in shaping user perception, usage behavior, and interaction experience. Thoughtful and well-executed design can promote the adoption and enhance the positive impact of these technologies, whereas poorly designed systems may lead to user discomfort, increased perceived risks in financial transactions, and hindered human-AI collaboration in financial scenarios. Thus, the design of AI applications in financial services remains an important area for further exploration and improvement.

On the other hand, the stability and security of the Metaverse present significant challenges in developing FinTech within virtual environments. Users' trust in platform stability and security is essential for their widespread adoption and sustained use in the shift from conventional financial channels to the Metaverse. Thus, technological challenges beyond establishing secure platforms encompass the critical task of cultivating and maintaining user trust in these systems (Siau & Wang, 2018). This requires not only robust security measures but also transparent interaction, reliable performance, and user-friendly interfaces that reinforce confidence in the platform's ability to safeguard sensitive financial activities and personal data. Another challenge is that current technologies are not advanced enough to provide fully immersive and comfortable experiences for users in the Metaverse. Enhancing the realism of virtual environments and the intelligence of conversational avatars is, therefore, essential to advancing user engagement and facilitating the growth of financial activities in these digital spaces. The potential of artificial intelligence, such as GenAI, to enhance the creation and adaptivity of the Metaverse is particularly promising and worthy of exploring.

A three-phase strategy is proposed to systematically address the identified technical challenges of FinTech digital transformation step by step. In the first phase, a comprehensive assessment of the technological landscape is necessary for researchers and practitioners to identify both the challenges and opportunities emerging in the transition toward the era of Industry 5.0. Building upon the insights from phase one, phase two should focus on improving the accuracy of AI technologies and refining their interface designs, alongside enhancing security, stability, and immersion of Metaverse. These technological improvements will be critical in fostering user trust, improving user experience, and promoting effective human-AI collaboration in the provision of financial services in both the physical world and the Metaverse. In the final phase, research should shift towards designing adaptive technological frameworks capable of evolving in response to users' changing needs, ensuring that the functions of platforms remain flexible and responsive as AI technologies and the Metaverse continue to develop.

4.2 Legal Perspective

On the legal front, the clear delineation of intellectual property rights has become an essential task as AI technologies gain increasing intelligence and functionality in the financial domain, particularly in the context of human-AI collaboration. Financial institutions must navigate issues related to copyright, patent rights, and trade secrets to ensure they can effectively leverage AI technologies while protecting their intellectual assets. Another legal challenge pertains to data privacy, as financial institutions increasingly collect and utilize sensitive customer information to provide users with personalized services. Striking a balance between offering personalized financial services and safeguarding individual privacy remains a persistent and evolving challenge for regulators and financial institutions.

To develop FinTech in the Metaverse, a lack of clear regulation frameworks governing virtual financial transactions is undoubtedly a key hindrance for financial institutions and investors. The dynamic evolution and virtual nature of financial activities in the Metaverse underscore the necessity for well-defined guidelines encompassing various aspects, like virtual assets, cross-border transactions, and contract enforcement. The inadequacy of legal regulations will increase the risks for financial institutions and investors, diminishing their willingness and confidence to participate in Metaverse-based financial activities. Therefore, adaptive and forward-looking legal structures to support financial activities within the Metaverse are urgent. In addition, the integration of GenAI into the Metaverse exacerbates concerns over data privacy, as vast amounts of personal and financial data would be processed within virtual worlds. Financial institutions need to prioritize developing transparent data-handling practices that respect user privacy while leveraging AI capabilities to enhance service delivery.

Thus, from the legal perspective, in the initial phase, a thorough legal assessment is necessary to identify the gaps in existing regulations and propose a coherent framework for the governance of financial activities in the Metaverse. This phase should focus on guiding the development of a general legal infrastructure that includes critical issues such as digital financial asset protection, virtual asset ownership, intellectual property, and contract enforcement. Based on the outcomes of this assessment, phase two should focus on reinforcing user protection by developing legal tools and policies to provide robust safeguards for personal data privacy. In the final phase, research should focus on creating adaptable legal frameworks that can evolve alongside technological advancements, ensuring that legal systems remain responsive to new challenges.

4.3 Ethical Perspective

From an ethical perspective, while AI technologies and Metaverse present exciting opportunities for enhancing financial services, it is crucial to ensure that these innovations do not exacerbate existing socioeconomic inequalities (Qian et al., 2024). One of the ethical considerations is providing equitable access to digital financial services, especially within the Metaverse. This includes addressing potential disparities in access to technologies and ensuring that all users, regardless of socioeconomic status, can benefit from these advancements. Another critical ethical concern is algorithmic biases (Siau & Wang, 2020). The use of AI-driven decision-making systems raises significant issues related to transparency and accountability. Financial institutions must implement rigorous monitoring and evaluation processes to ensure that AI algorithms do not produce biased outcomes or unintended consequences.

In addition, as financial institutions adopt more sophisticated AI systems, it is essential to consider their impact on employment and employee development. The digital transformation in the financial industry can lead to significant changes in job roles and responsibilities, potentially displacing certain occupations, altering certain job content, and replacing some jobs. This shift raises concerns about rising unemployment, particularly for workers engaged in routine tasks who may lack the necessary skills to adapt to new technologies (Siau, 2018). Therefore, it is crucial to ensure that technology development and adoption are human-centric, focusing on enhancing human labor instead of displacing it. Training and reskilling initiatives that equip employees with the skills necessary to collaborate effectively with AI technologies are worthy of investigation to help humans thrive in an AI-driven environment (Hyder et al., 2019).

In addressing these ethical challenges, the initial phase of research should focus on conducting a comprehensive assessment to identify potential gaps and barriers that may prevent equitable access to digital financial services and lead to biased outcomes. Phase two should concentrate on mitigation strategies to address identified ethical concerns, devoting to developing inclusive and fair financial services, minimizing AI bias, and emphasizing transparency and accountability. This phase should also include the development of effective initiatives for training and reskilling employees to adapt to the changing job landscape, thus fostering a collaborative relationship between humans and AI. In the final phase, research should shift to adaptation, developing flexible ethical frameworks that can evolve with advancements in the Metaverse and AI technologies. This phase should involve continuous monitoring and updating of ethical practices to address new and emerging challenges, ensuring that financial practices remain ethical and responsive to the evolving societal context and maintaining positive social impact.

5 Conclusions

FinTech digital transformation is poised to revolutionize the traditional financial industry. The rapid advancements in AI technologies (e.g., GenAI and humanoid robots) and the emergence of Metaverse present significant opportunities for promoting FinTech digital transformation. The integration of AI technologies into financial service delivery and the expansion of financial activities within virtual environments and the Metaverse are becoming increasingly critical trends. The new industrial paradigm of Industry 5.0 places a strong emphasis on human-centric development, underscoring the importance of human-AI collaboration throughout this process. Initial efforts to unlock the potential of advanced AI technologies and the Metaverse in FinTech digital transformation have been undertaken by various stakeholders, including governments, financial institutions, and technology companies. Nevertheless, challenges and obstacles remain. This study makes contributions by providing insights to promote the deeper integration of AI technologies and the Metaverse into the FinTech landscape, thereby advancing the progress of FinTech digital transformation. Through a systematic review of the existing literature and an analysis of current developments in AI technologies and the Metaverse in financial scenarios, we propose research opportunities and future directions from the perspectives of law, technology, and ethics, respectively.

Disclosure of Interests. The authors have no competing interests to declare that are relevant to the content of this article.

References

Adel, A.: Future of industry 5.0 in society: human-centric solutions, challenges and prospective research areas. J. Cloud Comput. **11**(1), 40 (2022)

Ante, L., Wazinski, F.P., Saggu, A.: Digital real estate in the Metaverse: an empirical analysis of retail investor motivations. Financ. Res. Lett. **58**, 104299 (2023)

Arner, D.W., Barberis, J., Buckley, R.P.: The evolution of Fintech: a new post-crisis paradigm. Geo. J. Int. Law **47**, 1271 (2015)

Bahoo, S., Cucculelli, M., Goga, X., Mondolo, J.: Artificial intelligence in Finance: a comprehensive review through bibliometric and content analysis. SN Bus. Econ. **4**(2), 23 (2024)

Belanche, D., Casaló, L.V., Flavián, C.: Artificial Intelligence in FinTech: UNDERSTANDING robo-advisors adoption among customers. Ind. Manag. Data Syst. **119**(7), 1411–1430 (2019)

Bharadwaj, A., El Sawy, O.A., Pavlou, P.A., Venkatraman, N.V.: Digital business strategy: toward a next generation of insights. MIS Q. **37**(2), 471–482 (2013)

Bhat, J.R., AlQahtani, S.A., Nekovee, M.: FinTech enablers, use cases, and role of future Internet of things. J. King Saud Univ. Comput. Inf. Sci. **35**(1), 87–101 (2023)

Birch, D.G., Richardson, V.J.: Metamoney: payments in the Metaverse. J. Payments Strategy Syst. **17**(2), 130–141 (2023)

Brunetti, F., Matt, D.T., Bonfanti, A., De Longhi, A., Pedrini, G., Orzes, G.: Digital transformation challenges: strategies emerging from a multi-stakeholder approach. TQM J. **32**(4), 697–724 (2020)

Buhalis, D., Leung, D., Lin, M.: Metaverse as a disruptive technology revolutionising tourism management and marketing. Tour. Manage. **97**, 104724 (2023)

Cao, L.: Ai4tech: X-AI enabling X-Tech with human-like, generative, decentralized, humanoid and Metaverse AI. Int. J. Data Sci. Analyt. 1–20 (2024)

Chen, Q.: Challenges and opportunities of Fintech innovation for traditional financial institutions. Front. Bus. Econ. Manage. **13**(3), 28–33 (2024)

Duan, H., Li, J., Fan, S., Lin, Z., Wu, X., Cai, W.: Metaverse for social good: a university campus prototype. In: Proceedings of the 29th ACM International Conference on Multimedia, pp. 153–161 (2021)

Fadhel, M.A., et al.: Navigating the Metaverse: unraveling the impact of artificial intelligence – a comprehensive review and gap analysis. Artif. Intell. Rev. **57**(9), 264 (2024)

Far, S.B., Rad, A.I., Asaar, M.R.: Blockchain and its derived technologies shape the future generation of digital businesses: a focus on decentralized finance and the Metaverse. Data Sci. Manage. **6**(3), 183–197 (2023)

Fitzgerald, M., Kruschwitz, N., Bonnet, D., Welch, M.: Embracing digital technology: a new strategic imperative. MIT Sloan Manag. Rev. **55**(2), 1 (2014)

Ford, T.J., et al.: Taking modern psychiatry into the Metaverse: integrating augmented, virtual, and mixed reality technologies into psychiatric care. Front. Digit. Health **5**, 1146806 (2023)

Fortune Business Insight. Global FinTech Market Size, Share & Trends Analysis, [2030] (2023). https://www.fortunebusinessinsights.com/fintech-market-108641

Ghobakhloo, M., Iranmanesh, M., Morales, M.E., Nilashi, M., Amran, A.: Actions and approaches for enabling Industry 5.0-driven sustainable industrial transformation: a strategy roadmap. Corp. Soc. Responsib. Environ. Manage. **30**(3), 1473–1494 (2023)

Gobble, M.M.: Digital strategy and digital transformation. Res. Technol. Manag. **61**(5), 66–71 (2018)

Hennig-Thurau, T., Aliman, D.N., Herting, A.M., Cziehso, G.P., Linder, M., Kübler, R.V.: Social interactions in the Metaverse: framework, initial evidence, and research roadmap. J. Acad. Mark. Sci. **51**(4), 889–913 (2023)

Huang, H., et al.: How economic systems enable Metaverse: value circulation in Metaverse via Web3. In: From Blockchain to Web3 & Metaverse, pp. 115–151. Springer (2023)

Hyder, Z., Siau, K., Nah, F.: Artificial intelligence, machine learning, and autonomous technologies in mining industry. J. Database Manage. **30**(2), 67–79 (2019)

Imerman, M.B., Fabozzi, F.J.: Cashing in on innovation: a taxonomy of FinTech. J. Asset Manag. **21**, 167–177 (2020)

Jovanović, A., Milosavljević, A.: VoRtex Metaverse platform for gamified collaborative learning. Electronics **11**(3), 317 (2022)

Jung, D., Glaser, F., Köpplin, W.: Robo-advisory: opportunities and risks for the future of financial advisory. In: Advances in Consulting Research: Recent Findings and Practical Cases, pp. 405–427 (2019)

Kye, B., Han, N., Kim, E., Park, Y., Jo, S.: Educational applications of Metaverse: possibilities and limitations. J. Educ. Eval. Health Prof. **18**, 32 (2021)

Lee, L.H., et al.: All one needs to know about Metaverse: a complete survey on technological singularity, virtual ecosystem, and research agenda (2021). arXiv:2110.05352

Mystakidis, S.: Metaverse. Encyclopedia **2**(1), 486–497 (2022)

Nah, F., Zheng, R., Cai, J., Siau, K., Chen, L.: Generative AI and ChatGPT: applications, challenges, and AI-human collaboration. J. Inf. Technol. Case Appl. Res. **25**(3), 277–304 (2023)

Nambisan, S., Wright, M., Feldman, M.: The digital transformation of innovation and entrepreneurship: progress, challenges and key themes. Res. Policy **48**(8), 103773 (2019)

Ng, D.T.K.: What is the Metaverse? Definitions, technologies and the community of inquiry. Australas. J. Educ. Technol. **38**(4), 190–205 (2022)

Obaigbena, A., Lottu, O.A., Ugwuanyi, E.D., Jacks, B.S., Sodiya, E.O., Daraojimba, O.D.: AI and human-robot interaction: a review of recent advances and challenges. GSC Adv. Res. Rev. **18**(2), 321–330 (2024)

Ooi, K.B., et al.: Banking in the Metaverse: a new frontier for financial institutions. Int. J. Bank Mark. **41**(7), 1829–1846 (2023)

Popescu, G.H., Valaskova, K., Horak, J.: Augmented reality shopping experiences, retail business analytics, and machine vision algorithms in the virtual economy of the Metaverse. J. Self-Govern. Manage. Econ. **10**(2), 67–81 (2022)

Qian, Y., Siau, K., Nah, F.: Societal impacts of artificial intelligence: ethics, legal, and governance issues. Societal Impacts **3**, 100040 (2024)

Raja Santhi, A., Muthuswamy, P.: Industry 5.0 or industry 4.0 S? Introduction to industry 4.0 and a peek into the prospective industry 5.0 technologies. Int. J. Interact. Des. Manuf. **17**(2), 947–979 (2023)

Riasanow, T., Flötgen, R.J., Setzke, D.S., Böhm, M., Krcmar, H.: The generic ecosystem and innovation patterns of the digital transformation in the financial industry (2018)

Ridzuan, N.N., Masri, M., Anshari, M., Fitriyani, N.L., Syafrudin, M.: AI in the financial sector: the line between innovation. Regul. Ethical Respons. Inf. **15**(8), 432 (2024)

Sabir, A.A., Ahmad, I., Ahmad, H., Rafiq, M., Khan, M.A., Noreen, N.: Consumer acceptance and adoption of AI robo-advisors in fintech industry. Mathematics **11**(6), 1311 (2023)

Saivasan, R., Lokhande, M.: Exploring use cases of generative AI and Metaverse in financial analytics: unveiling the synergies of advanced technologies. Int. J. Glob. Bus. Compet. 1–10 (2023)

Şanlısoy, S., Çiloğlu, T.: A view of the future of the Metaverse economy on the basis of the global financial system: new opportunities and risks. J. Corp. Govern. Insur. Risk Manage. **10**(1), 28–41 (2023)

Sautunnida, L., Zain, N.R.M., Zakri, I.M.M., Yahya, A.: The application of artificial intelligence in Metaverse: a new challenge on personal data protection in the financial system. In: Islamic Sustainable Finance, Law and Innovation: Opportunities and Challenges, pp. 117–126. Springer Nature Switzerland, Cham (2023)

Schiller, S., Nah, F.F.H., Luse, A., Siau, K.: Men are from Mars and women are from Venus: dyadic collaboration in the Metaverse. Internet Res. **34**(1), 149–173 (2024)

Shneiderman, B.: Human-centered artificial intelligence: three fresh ideas. AIS Trans. Hum. Comput. Interact. **12**(3), 109–124 (2020)

Siau, K.: Interorganizational systems and competitive advantages – lessons from history. J. Comput. Inf. Syst. **44**(1), 33–39 (2003)

Siau, K.: Education in the age of artificial intelligence: how will technology shape learning? Global Analyst **7**(3), 22–24 (2018)

Siau, K., et al.: FinTech empowerment: data science, artificial intelligence, and machine learning. Cutter Bus. Technol. J. **31**(11/12), 12–18 (2018)

Siau, K., Wang, W.: Building trust in artificial intelligence, machine learning, and robotics. Cutter Bus. Technol. J. **31**(2), 47–53 (2018)

Siau, K., Wang, W.: Artificial intelligence (AI) ethics – ethics of AI and ethical AI. J. Database Manage. **31**(2), 74–87 (2020)

Soomro, Z.A., Ali, Q., Parveen, S.: Diffusion of Industry 5.0 in the financial sector: a developmental study. In: Proceedings of the BAM (2022)

Stephanidis, C., et al.: Seven HCI grand challenges. Int. J. Hum. Comput. Interact. **35**(14), 1229–1269 (2019)

Swami, P.: Metaverse: Transforming the User Experience in the Gaming and Entertainment Industry. Research, Innovation, and Industry Impacts of the Metaverse, pp. 115–128 (2024)

Wang, W., Siau, K.: Artificial intelligence, machine learning, automation, robotics, future of work, and future of humanity – a review and research agenda. J. Database Manage. **30**(1), 61–79 (2019)

Wang, Y., Wang, L., Siau, K.: Human-centered interaction in virtual worlds: a new era of generative artificial intelligence and Metaverse. Int. J. Hum. Comput. Interact. (2024)

Wirtz, J., Pitardi, V.: How intelligent automation, service robots, and AI will reshape service products and their delivery. Ital. J. Mark. **2023**(3), 289–300 (2023)

Yang, Y., Siau, K., Xie, W., Sun, Y.: Smart health: intelligent healthcare systems in the Metaverse, artificial intelligence, and data science era. J. Org. End User Comput. **34**(1), 1–14 (2022)

Artificial Intelligence (AI) and Virtual Reality Convergence in Financial Services: The Power of Digital Twin Robo-Advisers

Marco I. Bonelli[1]([✉]) [iD] and Jiahao Liu[2] [iD]

[1] "Ca'Foscari" Dept. of Management, University of Venice, Venice, Italy
mibonelli6@gmail.com
[2] Judge Business School, University of Cambridge, Cambridge, UK

Abstract. The financial services sector has undergone significant transformation due to technological advancements, globalization, and evolving consumer demands. Fintech innovations have revolutionized financial interactions, democratizing access and reducing costs. Artificial Intelligence (AI), particularly in robo-advisers, has emerged as a key driver of this change, automating investment management. Meanwhile, Virtual Reality (VR) technologies are enhancing user engagement by integrating physical and virtual experiences, transforming professional practices, learning, and entertainment. By combining AI-powered robo-advisers with Digital Twin (DT) capabilities and VR, this research explores how users can interact with their financial data in immersive 3D environments. These enhanced robo-advisers allow users to visualize and simulate financial scenarios, making complex information more accessible and decision-making more informed. The integration of VR offers a highly interactive, immersive experience that elevates personal financial management. This paper highlights the transformative potential of AI and VR in financial services, with the Digital Twin robo-adviser representing the convergence of both technologies. This convergence enables users to engage with their finances in ways that set these DT-enhanced robo-advisers apart from other solutions. The result is a new standard for personalized, efficient, and engaging advisory services that maximize opportunities for both users and service providers.

Keywords: Virtual Reality (VR) · Augmented Reality (AR) · Fintech · Robo-Advisers · Digital Twin (DT) · Financial Services · Artificial Intelligence (AI) · Financial Technology Integration

1 Introduction

1.1 Evolution of the Nature of "Financial Services"

The financial services sector has significantly transformed due to technological advancements, global economic shifts, and evolving consumer expectations. Aggarwal and Karwasra [1] highlight that the revolution in software technology and globalization, driven

K.-W. Huang et al. (Eds.): ICFT 2024, CCIS 2437, pp. 137–150, 2025.
https://doi.org/10.1007/978-981-96-3811-6_13

by smart devices and enhanced internet connectivity, has been pivotal in creating a digital economy. This shift has fostered the rise of fintech firms, reshaping financial services globally.

Fintech, or the integration of technology into financial services, has created new paradigms for businesses and consumers.

Meena [2] analyzes how fintech companies, offering services like online banking, digital wallets, and peer-to- peer lending, have democratized access to financial services. These innovations have brought efficiency and cost reductions, catering to underserved communities with faster, cheaper, and more convenient alternatives to traditional banking.

Historically, the evolution of financial services aligns with advancements in computing technology. Radianti et al. [3] discusse the early focus on education and research in user services and their adaptation to evolving computer environments. This foundation was crucial for the development of today's sophisticated financial services. The transition from traditional banking to a technology-driven sector reflects the interplay between technological advancements, cultural influences, and economic reforms. The move towards a digital economy, and the rise of fintech firms have reshaped the financial services landscape, presenting new opportunities and challenges for businesses and consumers alike.

1.2   Artificial Intelligence (AI) in the Financial Services Industry

Artificial intelligence (AI) refers to the development of computer systems capable of performing tasks that typically require human intelligence, such as decision-making, problem-solving, and language understanding. Rooted in computer science and philosophy, AI has rapidly evolved and is now widely used in everyday life, from virtual assistants to automated industrial processes. Its applications include natural language processing, robotic automation, and machine learning, which enable machines to simulate intelligent behaviour [4].

Generative AI takes AI a step further by enabling the creation of new content, such as images, text, and even software, without requiring technical expertise. Unlike traditional AI, which powers recommendation systems like Google or Netflix, generative AI learns patterns from existing data and generates entirely new outputs. For example, after learning the features of a cat, generative AI can produce multiple variations of cat images. This distinguishes it from discriminative AI, which classifies existing data rather than creating new content. The use of generative AI has expanded from text and image generation to coding, audio, video, and virtual environments, transforming industries and offering vast creative potential [5]. Tech companies like Microsoft, IBM, and Google have heavily invested in AI research and development, with billions flowing into AI startups each year. In the financial sector, AI adoption is particularly high, with machine learning widely used since the late 2000s and generative AI gaining traction. In 2023, financial services invested around $35 billion in AI, with banking leading the way at $21 billion. Top adopters include Capital One, JPMorgan Chase, and the Royal Bank of Canada [6]. AI has become integral across financial institutions, especially in operations, risk management, and customer service. However, challenges like data privacy and talent shortages continue to hinder broader adoption. In 2023, over two-thirds of financial institutions used AI for data analytics, while more than 40% explored generative AI. The global generative AI market in finance is expected to grow at 28.1% annually from 2023 to 2032, potentially adding $200 to $340 billion in value to the banking sector, equivalent to up to 5% of industry revenue [6].

1.3 Evolution and Foundation of AR and VR, Technology and User Experience

Augmented Reality (AR) and Virtual Reality (VR) have revolutionized how we interact with digital content, merging the physical and virtual worlds in innovative ways. Cipresso et al.[7] trace the evolution of VR and AR from their origins in the 1960s to recent advancements, highlighting their potential impact on sectors like education, healthcare, and entertainment. AR overlays digital information onto the real world, allowing users to engage with both simultaneously, while VR immerses users in a fully digital environment, isolating them from the physical world. This distinction shapes their respective applications in educational and professional settings.

Christensen et al.[8] discuss the transition of VR and AR from niche innovations to significant technological milestones, predicting their integration into various industries due to their immersive, interactive capabilities. The intersection of AR and VR with the arts exemplifies their transformative potential.

Ghazwani and Smith [9] highlight the significance of user experience in AR, emphasizing the challenges of creating intuitive interfaces for virtual content. AR and VR differ technologically and in user experience, requiring distinct design approaches.

Chandana et all.[10] identify key UX design principles for AR and VR, such as immersion and spatial awareness, crucial for engaging user experiences. Designers must address issues like motion sickness, field of view limitations, and cognitive load to ensure usability. Ahmeth [11] discusses the evolution of AR and VR into Mixed Reality (MR) and the integration of Artificial Intelligence (AI), creating environments where physical and digital objects coexist and interact. This technological convergence promises more authentic and emotionally resonant experiences. There are numerous books on artificial intelligence, particularly in its application to finance. Zhang and Zhao, in *Artificial Financial Intelligence in China* [12], demonstrate the significant benefits of AI technologies like facial, speech, and semantic recognition. Barrau and Douady, in *Artificial Intelligence for Financial Markets* [13], present the AI technique of polymodels, applying it to stock return predictions. Anshari et al. [14] explore the use of robo-advisors with digital twin capabilities for personal financial management, showcasing the expanding role of AI in financial services.

Understanding AR and VR from both technological and user experience perspectives reveals their capabilities and limitations. As these technologies progress, they promise to transform interactions with digital content and perceptions of the world.

1.4 Types of Robo-Advisers

Robo-advisors are AI-driven platforms that automatically develop and oversee investment portfolios for users, providing a cost-effective alternative to traditional human financial advisors [15].

A. **Full-Scale Optimization:** Robo-advisors initially adopted the Modern Portfolio Theory (MPT) due to its reliable and time-tested approach based on Nobel Prize-winning research. MPT uses mean-variance analysis to maximize expected returns while minimizing risk, appealing to investors seeking automated solutions. However, MPT has limitations, such as over-reliance on data inputs and the assumption that asset returns are normally distributed [16].

B. **Digital Twin:** This cutting-edge concept involves creating a digital replica of a user's environment, enabling robo-advisors to provide highly personalized and real-time financial advice. By continuously collecting data from a client's virtual environment, digital twins offer tailored recommendations based on life changes [17]. The integration of digital twins could revolutionize the advisory experience by making it more interactive and responsive to individual needs [18].

C. **Natural Language Processing (NLP):** NLP technology aims to address one of the robo-advisory industry's key challenges: the lack of emotional connection with clients. AI chatbots utilizing NLP can simulate human-like conversations, building trust and enhancing client relationships. This technology can improve the user experience, foster customer loyalty, and encourage adherencetoinvestmentadvice. In summary, robo-advisors are evolving by integrating advanced investment strategies, personalized digital twin technologies, and NLP to enhance user experience and client engagement. These innovations address the limitations of traditional methods and offer new ways to meet clients' diverse needs [19].

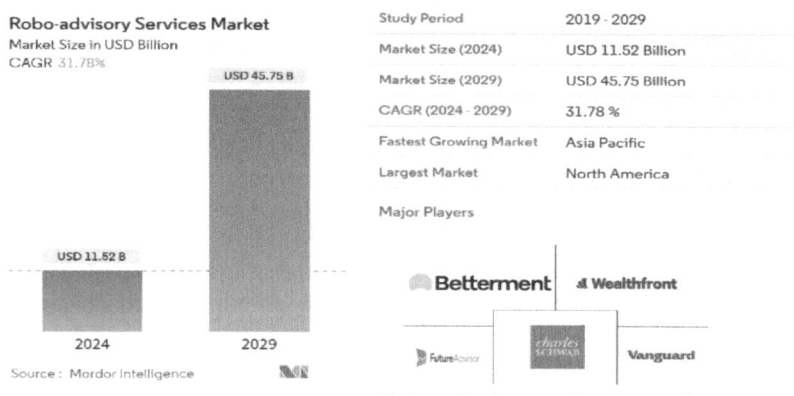

Fig. 1. Robo-advisory global market [20]

The robo-advisory services market is forecasted to expand from USD 11.52 billion in 2024 to USD 45.75 billion by 2029, growing at a CAGR of 31.78% during this period. The swift digitization of the financial sector has greatly boosted digital investments, with robo-advisors being a key factor in this transformation [20] (Fig. 1).

1.5 Robo-Advisers' Development

As the market becomes saturated, robo-advisors must differentiate themselves to attract clients and retain loyalty, which drives the evolution of new investment strategies [21]. Initially, many relied on Modern Portfolio Theory (MPT), a Nobel Prize-winning approach designed to maximize returns while minimizing risk through mean-variance analysis [22].

MPT's mathematical model involves balancing expected returns and risks, but its limitations have become more evident as the robo-advisory landscape evolves. Problems include heavy reliance on data, over-concentration in certain assets, and assumptions that asset returns follow normal distributions, which may not always hold in volatile markets.

Rather than abandoning MPT, many firms have introduced enhancements through a framework called "Multidimensional Improvement of Modern Portfolio Theory" (MPT+). This updated model adds more factors and methods, making portfolios more flexible and better equipped to handle real-world risks [23].

New techniques like Full-Scale Optimization, Risk Parity, and Scenario Optimization are now being adopted. For example, Schwab Intelligent Portfolios have implemented Full-Scale Optimization, showing how these advanced strategies can attract more client investments [24].

In short, as the robo-advisory field advances, models like MPT+ and new optimization strategies are playing a key role in improving investment outcomes and meeting the diverse needs of investors [14].

1.6 How Artificial Intelligence is Impacting Robo-Advising

Artificial intelligence (AI) is anticipated to play a transformative role in robo-advising. Grennan & Michaely [25] found that 57% of the 290 FinTech firms they studied use AI to generate investment signals. With the integration of AI, banks can provide chatbots, virtual assistants, and robo-advisors that directly interact with clients, offering a range of services [26]. AI's ability to process massive datasets enables it to extract insights that can inform more effective investment strategies. By utilizing machine learning, robo-advisors can adapt to evolving market conditions, providing more accurate forecasts and data-driven investment decisions. AI also allows robo-advisors to offer more personalized recommendations by considering factors such as individual preferences, risk tolerance, and financial goals. This enables the creation of customized portfolios tailored to each investor's specific needs. Furthermore, AI-powered robo-advisors can leverage natural language processing to improve communication with investors, responding to their queries and offering guidance in a more conversational and intuitive manner. Another key benefit of AI in robo-advising is its ability to enhance risk management. AI algorithms can monitor market trends and economic indicators in real time, adjusting investment strategies to mitigate potential risks. This real-time assessment can help maximize returns and minimize losses. Additionally, AI can incorporate insights from behavioral finance, accounting for emotional factors that influence investor decisions, leading to strategies that align better with investors' psychological tendencies and long-term goals. AI-driven automation streamlines various robo-advising processes, such as portfolio rebalancing, tax optimization, and trade execution, reducing human error and improving operational efficiency. This also helps lower costs and creates a smoother, more consistent user experience. Moreover, AI-powered systems can detect fraudulent activities and bolster cyber-security by analyzing patterns, anomalies, and historical data to protect investor information and ensure transaction integrity.Overall, the use of AI in robo-advising can significantly enhance decision-making, increase personalization, and improve the overall efficiency of the investment process. However, it is crucial to address ethical concerns, data privacy, and transparency, as the growing reliance on AI introduces new challenges for providers and regulators. Compared to human advisors, robo-advisors may deliver better returns due to the lower costs associated with AI-powered financial advisory services [27].

2 Methodology of the Study

This research is an exploratory study, focusing on the relatively new concept of digital twins and their application in robo-advisory services for personal finance management. Currently, there is limited research available in this specific area, providing a valuable opportunity for further investigation and exploration by researchers in the future [28]. Exploratory research is typically qualitative, offering a means to develop a deeper understanding of emerging topics [29].

In this study, secondary data collection was utilized to achieve the research objectives. This study highlights how digital twin robo-advisors represent a significant advancement, embodying an immersive, holistic virtual reality instrument that transforms financial advisory services into a significative virtual reality experience.

The research suggests a three-stage approach for conducting a literature review to develop a comprehensive framework. These stages are input, processing, and output.

The input stage addresses challenges related to identifying relevant literature and sources. Processing involves categorizing the literature and determining appropriate interpretation methods, such as cognitive or construct-level analysis, literary streams, and theoretical frameworks. Finally, the output stage involves writing the literature review and analyzing how the existing body of knowledge influences the proposed framework.

The three-stage literature review process begins with the inputs stage, where pertinent books, articles, and other relevant sources are identified. This step is crucial for building a foundation of existing knowledge and ensuring that the literature review is comprehensive and well- informed.

The processing stage follows, which involves organizing the identified literature into categories and determining how to interpret the information. This may include analyzing cognitive and construct-level perspectives, exploring different literary streams, and applying relevant theories to the research topic.

In the final output stage, the literature review is written, combining and evaluating the information collected earlier. This step involves a critical assessment of the literature and an explanation of how the existing knowledge influences the proposed framework. The aim is to present a clear, coherent narrative that underscores key findings and insights, showcasing the research's relevance and significance.

By following this structured approach, the research aims to provide a comprehensive and detailed literature review that serves as a foundation for future studies in the field of digital twins and robo-advisory services for personal finance management. This process not only enhances the understanding of the topic but also contributes to the development of a robust framework that can guide further research and practical applications in this emerging area. Furthermore, this study underscores the potential of digital twin technology to revolutionize robo-advisory services by creating an immersive and holistic virtual reality experience, significantly enhancing the way users interact with and manage their financial landscapes.

3 Virtual Reality in the AI Based Digital Twin Robo-Adviser

3.1 Twin Digital Technology: Applications and Impact

Digital Twin (DT) technology creates a virtual replica of a real-world entity, using real-time data and analytics to provide feedback, recommendations, and solutions [30]. It is widely used across various industries. In healthcare, digital twins help cardiovascular experts develop 3D models of the human heart for education and training [31]. These models also assist in diagnosis, monitoring, surgery, and drug development, enhancing patient safety and improving clinical outcomes [32].

In product and machine maintenance, digital twins leverage embedded sensors to monitor performance and predict issues before they occur, increasing operational efficiency [33]. The rise of technologies such as the Internet of Things (IoT), AI, and machine learning has further boosted productivity and innovation across sectors. Digital twins also play a role in urban planning by simulating city environments. For example, companies like 51 World have created 3D models of cities like Shanghai and Singapore, improving infrastructure design and traffic management systems [34].

In finance, digital twins enhance robo-advisors by collecting real-world data, simulating scenarios, and offering personalized recommendations. This improves decision-making and operational efficiency [35].

The Digital Twin market is projected to grow from USD 26.14 billion in 2024 to USD 125.89 billion by 2029, with a 36.94% CAGR. This surge is driven by the need for digital twin solutions that simplify manufacturing, optimize maintenance, monitor assets, reduce downtime, and create connected products (37). Technologies like IoT, cloud platforms, machine learning, and AI have played a key role in expanding digital twin use in industries such as manufacturing and engineering. The COVID-19 pandemic further accelerated the adoption of Industry 4.0 technologies, boosting market demand (38).

3.2 Robo- Advisers and Digital Twin Technology

When combined with DT, robo-advisers can transform from static platforms into dynamic and comprehensive financial advisory services, offering significantly more value for users' financial well-being. The integration of DT allows robo-advisers to monitor users' financial activities in real-time, providing timely notifications and recommendations based on a thorough analysis of their digital footprint [36]. This combination enhances the personalization and customization of the services provided, enabling users to make better-informed financial decisions and improve their overall financial management

The bibliographic analysis of pertinent work in the domain under investigation was conducted as follows:

The initial step involved conducting keyword searches across Scopus databases and Google Scholar, focusing on "Digital Twin." Most of the existing research on digital twins pertains to Industry 4.0, cyber-physical systems, machine learning, IoT, and smart manufacturing, with a limited exploration into FinTech with DT integration.

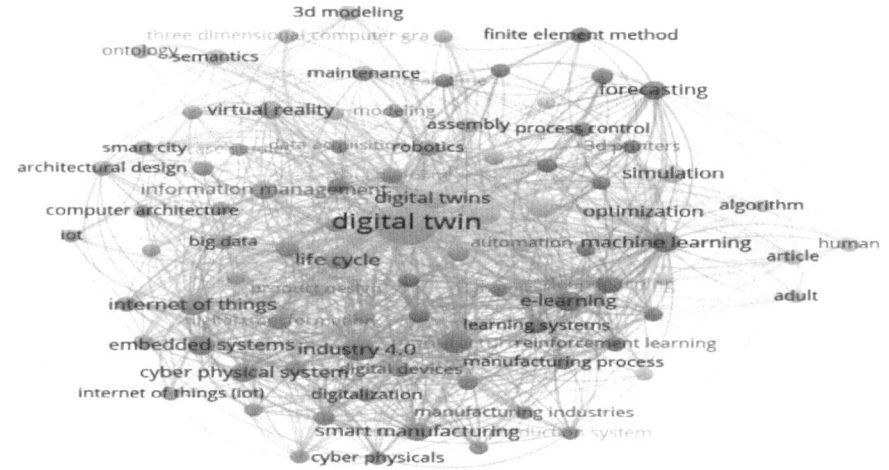

Fig. 2. Bibliographic analysis on digital twin [14]

This study aims to pioneer exploratory research on FinTech and robo-advisor trends. The proposed literature review process comprises three stages: input, processing, and output.

Figure 2, [14] illustrates a digital twin scenario augmenting the capabilities of robo-advisors in personal financial management. It underscores the interconnectedness and holistic nature of both physical and digital ecosystems, envisioning future scenarios for personal finance management. Within the context of a robo-advisor integrated with DT capabilities, personal data converges into a platform operating as a digital twin.

The evolution of financial robo-advisors with DT integration is propelled by widespread smartphone adoption. User activities are digitally monitored and recorded, offering insights for value generation. Smartphones have emerged as primary devices for various financial activities, encompassing e-commerce, digital transactions, investments, payments, and social media interactions [37] (Fig. 3).

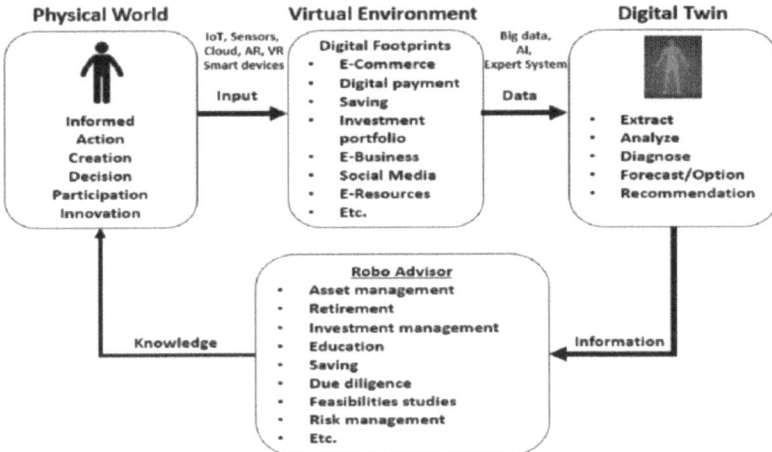

Fig. 3. [16] Integrating Digital Twin capabilities into robo- advisor software involves creating a digital replica of the user's environment. Data collected from this virtual environment would be used by the robo-advisor to provide tailored advice to the real user. In the scenario of a robo-advisor equipped with DT capabilities, personal data becomes integrated into a platform functioning as a digital twin [16].

Data from sources like IoT, sensors, cloud computing, and smart devices form a unique digital footprint of individuals' financial activities. This information is logged and analyzed to provide personalized forecasts and recommendations (14). Users can engage with their digital twin (DT), receiving tailored guidance on asset management, retirement planning, investments, and savings strategies, significantly enhancing their financial well-being (42).

Unlike traditional robo-advisors that use incomplete data, digital twins (DTs) provide a full financial picture by combining data from various sources. This allows for real-time predictions and personalized advice, offering better insights for managing assets and assessing risks [14].

The near-instant sync between physical and digital systems helps prevent issues and improve decision-making. Users can track their progress and make smarter financial choices. Integrating DTs into FinTech helps people evaluate options, boost performance, and achieve better financial results [14].

3.3 AI Based Digital Twin Investment Portfolio Simulation

Being the Digital Twin a virtual model or replica of something real, for investors it is like creating a virtual version of a portfolio or an asset. By analyzing this digital version, they can simulate and predict how the actual portfolio might perform under different conditions. It is like running tests without affecting the real money, so they can be ready to use this to make smarter decisions [38].

By creating a digital version of their portfolio (using data and complex algorithms), they can simulate how it might perform in the future. This helps them answer questions like, *"What will happen if the market crashes?"* or *"What if we take more risks?"*.

How does it work?

1. **Simulating different scenarios:** The Digital Twin can simulate how different investment strategies might perform in various market conditions. For example, if the market is booming or if it's facing a downturn.
2. **Answering 'What if?' questions:** This lets bankers test out different strategies and see which one gives the best return on investment (ROI).

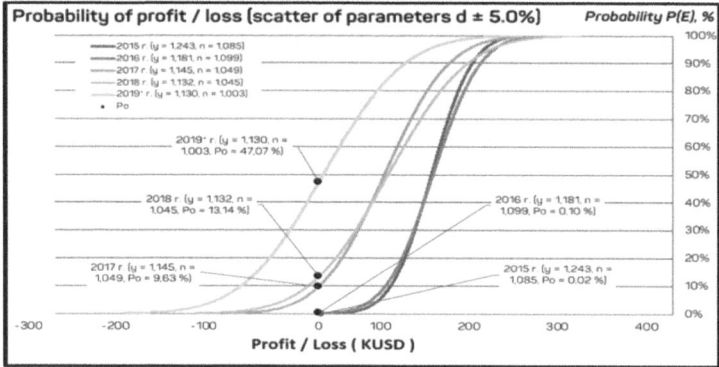

Fig. 4. [38]: Digital Twin Portfolio Simulation (Probability of Profit/Loss)

Example from Fig. 4:

1. **Good Performance (2015–2016):** In this case, the asset (let's say a stock) did well in 2015 and 2016. The risk of losing money was low, and the probability of making a profit and hitting a good **EBITDA** (a measure of profitability) was high. Since the asset didn't lose money, its capital reserves stayed the same.
2. **Declining Performance (2017–2019):** In 2017, things started going south. The asset lost value (about $100k), and the risk of losing more money went up. The asset had to dip into its capital reserves (like a savings account) to cover the losses. By 2019, the situation worsened due to poor management decisions, and the chance of making a profit or reaching the target EBITDA became even lower. The risk was higher [38].

Decisions to Make:
The Digital Twin simulation helps investors decide whether to:

- **Reevaluate** the asset: Should they keep or sell it?
- **Reinvest** in it: Should they put more money into it to try and recover?
- **Diversify**: Should they invest in other things to balance the losses?

For a diversified portfolio:
If the portfolio has multiple assets (stocks, bonds, etc.), the Digital Twin and data analytics can help find the best strategy for managing all of them. This includes deciding how to distribute dividends or profits and how to optimize turnover (how often they buy and sell). This would help increase overall returns [38].

3.4   Convergence of Virtual Reality and AI Based Twin Digital Technology

The convergence of Artificial Intelligence (AI)-powered Digital Twin (DT) technology and Virtual Reality (VR) marks a new frontier in robo-advisory services, offering an advanced and immersive financial management experience. This integration amplifies the capabilities of both technologies, allowing users to visualize, simulate, and interact with their investment portfolios in an unprecedented manner [14].AI-driven DT technology generates a highly accurate, real-time virtual replica of an investment portfolio or financial ecosystem. This virtual model continuously learns from new data and evolving market conditions, enabling it to simulate future performance and recommend optimal strategies for maximizing returns. The "what-if" scenarios powered by AI enable sophisticated modeling of market trends, risks, and potential investment outcomes, making financial planning and portfolio management more precise [39].

When coupled with VR, the AI-based Digital Twin creates an immersive 3D environment where users can virtually navigate their financial landscape. Instead of merely viewing numbers on a screen, they can step into a virtual simulation, where complex financial data is represented visually, making it easier to comprehend. Users can walk through their portfolio, test different investment strategies, and immediately see how these decisions affect their financial goals, such as retirement planning or asset allocation [39].

This convergence also enhances real-time decision-making. For example, when an AI-powered DT detects a market shift, it can update the virtual environment in real-time, enabling users to make swift, informed decisions. Additionally, the VR setting enables interactive financial literacy programs, where users can learn essential concepts such as risk management, budgeting, and long-term planning in a dynamic and engaging format [40].

The combined power of AI-based DT and VR offers a deeply personalized and interactive advisory service, tailored to each user's financial behavior, goals, and preferences. This technological fusion not only transforms financial management but also provides users with a tool that is highly adaptive to changing conditions and uniquely tailored to individual needs [41].

4   Conclusion

The convergence of AI-based Digital Twin technology and Virtual Reality (VR) in robo-advisory services represents a state-of-the-art advancement in financial management. This integration creates a highly immersive and interactive platform that allows users to engage with their financial data like never before. By combining AI's predictive power with the immersive nature of VR, robo-advisors now offer the most advanced toolset available to investors [42].

AI-powered Digital Twins provide real-time simulations of investment portfolios, allowing users to explore multiple scenarios, assess risks, and optimize their strategies with precision. The addition of VR creates a lifelike, 3D financial environment where users can visualize their portfolios, simulate various financial decisions, and gain deeper insights into how these choices impact their long-term financial goals. This cutting-edge approach not only simplifies complex data but makes financial management more intuitive, educational, and accessible [43].

This blend of AI and VR positions the AI-based Digital Twin robo-advisor as the ultimate solution for investors. By delivering a highly personalized, real-time experience, it outperforms traditional financial advisory methods. It equips users with the ability to foresee market conditions, make data-driven decisions, and better plan for the future, whether they are navigating short-term investments or long-term financial milestones like retirement.

In conclusion, a robo-advisor enhanced with AI and VR capabilities is the epitome of modern financial technology—offering a comprehensive, interactive, and personalized service that is unparalleled in its ability to help investors achieve their financial goals. With the integration of these technologies, robo-advisors are setting a new standard in financial services, ensuring users can make well-informed, strategic decisions and maximize their returns in an increasingly complex financial landscape.

Disclosure of Interests. The authors have no competing interests to declare that are relevant to the content of this article.

References

1. Aggarwal, V., Karwasra, N.: A bibliometric analysis on trade openness and economic growth: current dynamics and future direction. Competitiveness Rev. Int. Bus. J. (2023). https://doi.org/10.1108/CR-11-2022-0177
2. Meena, R.: The impact of financial technology on financial services: a comprehensive analysis. Res. Rev. Int. J. Multi. **8**(3), 185–192 (2023). https://doi.org/10.31305/rrijm.2023.v08.n03.022
3. Radianti, J., Majchrzak, T.A., Fromm, J., Wohlgenannt, I.: A systematic review of immersive virtual reality applications for higher education: design elements, lessons learned, and research agenda. Comput. Educ. **147**, 103778. ISSN 0360-1315 (2020). https://doi.org/10.1016/j.compedu.2019.103778
4. Thormundsson, B.: AI transformation timeframe for organizations worldwide. LinkedIn (2024). https://www.linkedin.com/pulse/ai-transformation-timeframe-organizations-worldwide. Accessed 13 Aug 2024
5. Marr, B.: Generative AI in Practice (2023). Amazon.in
6. Statista. AI transformation timeframe for organizations worldwide (2024). https://www.statista.com/statistics/1450526/ai-transformation-timeframe-organizations-worldwide/. Accessed 14 Aug 2024
7. Cipresso, P., Giglioli, I.A.C., Raya, M.A., Riva, G.: The past, present, and future of virtual and augmented reality research: a network and cluster analysis of the literature. Front. Psychol. 2086 (2018). https://doi.org/10.3389/fpsyg.2018.02086
8. Christensen, C.M., McDonald, R., Altman, E.J., Palmer, J.E.: Disruptive innovation: an intellectual history and directions for future research. J. Manage. Stud. **55**(7), 1043–1078 (2018). https://doi.org/10.1111/joms.12349
9. Ghazwani, Y., Smith, S.: Interaction in augmented reality: challenges to enhance user experience. In: Proceedings of the 2020 4th International Conference on Virtual and Augmented Reality Simulations, pp. 39–44 (2020). https://doi.org/10.1145/3385378.3385384

10. Chandana, B.H., Shaik, N., Chitralingappa, P.: Exploring the frontiers of user experience design: vr, ar, and the future of interaction. In 2023 International Conference on Computer Science and Emerging Technologies (CSET), pp. 1–6. IEEE (2023). https://doi.org/10.1109/CSET58993.2023.10346724

11. Ahmet, E.F.E.: Taking virtual reality and augmented reality to the next level: artificial intelligence with mixed reality. Kamu Yönetimi ve Teknoloji Dergisi 4(2), 141–165 (2022). https://doi.org/10.58307/kaytek.1185712

12. Zhao, D., Zhang, W.: Intelligent finance wealth management in China. Artif. Financ. Intell. Chin 129–146. Singapore: Springer (2021). https://doi.org/10.1007/978-981-16-5592-0_7

13. Barrau, T., Douady, R.: Artificial intelligence for financial markets. Springer Cham (2022). https://doi.org/10.1007/978-3-030-97319-3

14. Anshari, M., Almunawar, M.N., Masri, M.: Digital twin: financial technology's next frontier of robo-advisor. J. Risk Financ. Manage. 15(4), 163 (2022). https://doi.org/10.3390/jrfm15040163

15. Abraham, F., Schmukler, S.L., Tessada, J.: Robo-advisors: investing through machines. World Bank Res. Policy Briefs 134881 (2019)

16. Bonelli, M.I., Liu J. Revolutionizing Robo-Advisors: Unveiling Global Financial Markets, AI-Driven Innovations and Technological Landscapes for Enhanced Investment Decisions. Adv. Sci. Technol. Eng. Syst. J. 9(2), 33–44 (2024). https://doi.org/10.25046/aj090205

17. Enyedy, N., Yoon, S.: Immersive environments: learning in augmented + virtual reality. In U. Cress, C. Rosé, A. F. Wise, & J. Oshima (eds.), International Handbook of Computer-Supported Collaborative Learning, pp. 389–405. Springer International Publishing (2021). https://doi.org/10.1007/978-3-030-65291-3_21

18. Gervasi, O., Perri, D., Simonetti, M.: Strategies and system implementations for secure electronic written exams. IEEE Access 1 (2022)

19. Bonelli, M.I., Döngül, E.S.: Robo-advisors in the financial services industry: recommendations for full-scale optimization, digital twin integration, and leveraging natural language processing trends. In: 2023 9th International Conference on Virtual Reality (ICVR), pp. 268–275.Xianyang, China (2023). https://doi.org/10.1109/ICVR57957.2023.10169615

20. Mordor Intelligence, Digital Twin Market Size Report on Industry Size & Market Share Analysis Growth Trends & Forecasts (2024–2029) Source (2023). https://www.mordorintelligence.com/industry- reports/digital-twin-market. Accessed 1 Jun 2024

21. Research, Markets Robo-advisory Services Market Growth, Trends, COVID-19 Impact, and Forecasts (2022–2027) (2022). https://www.researchandmarkets.com/reports/5120011/roboadvisoryservices-market-growth-trends. Accessed 18 Jan 2023

22. Michaud, R.O.: The markowitz optimization enigma: is 'optimized' optimal? Financ. Anal. J. 45, 31–42 (1989). https://doi.org/10.2469/faj.v45.n1.31

23. Best, M., Grauer, R.: On the sensitivity of mean-variance-efficient portfolios to changes in asset means: some analytical and computational results. Rev. Financ. Stud. 4, 315–342 (1991). https://doi.org/10.1093/rfs/4.2.315

24. Beketov, M., Lehmann, K., Wittke, M.: Robo advisors: quantitative methods inside the robots. J. Asset Manag. 19, 363–370 (2018). https://doi.org/10.1057/s41260-018-0092-9

25. Grennan, J., Michaely, R.: Fintechs and the market for financial analysis. J. Financ. Quant. Anal. 56(6), 1877–1907 (2021). https://doi.org/10.1017/S0022109020000721

26. Singh, N., Singh, D.: Chatbots and virtual assistant in indian banks. Industrija 47, 75–101 (2019). https://doi.org/10.5937/industrija47-24578

27. Brenner, L., Meyll, T.: Robo-advisors: a substitute for human financial advice?. J. Behav. Exp. Financ. 25, 100275. ISSN 2214-6350 (2020). https://doi.org/10.1016/j.jbef.2020.100275

28. Cooper, D.R., Schindler, P.S., Sun, J.: Business research methods. New York: Mcgraw-Hill 9, 1–744 (2006)

29. Zikmund, W.G., Carr, J.C., Griffin, M.: Business Research Methods (Book Only). Cengage Learning, Melbourne (2013)
30. Mourtzis, D.: Advances in adaptive scheduling in industry 4.0. Front. Manuf. Technol. **2**, 937889 (2022). https://doi.org/10.3389/fmtec.2022
31. Bruynseels, K., Santoni de Sio, F., van den Hoven, J.: Digital twins in health care: ethical implications of an emerging engineering paradigm. Front. Genet. **9**, 31. PMID: 29487613; PMCID: PMC5816748 (2021). https://doi.org/10.3389/fgene.2018.00031
32. Orcajo, Enrique Morales. Digital twin applications in healthcare—the revolution of the next decade. Linkedin (2021). https://www.linkedin.com/pulse/6-digital-twin-applicationshealth care-revolution-enrique. Accessed 3 Jun. 2024
33. Perno, M., Hvam, L., Haug, A.: Implementation of digital twins in the process industry: a systematic literature review of enablers and barriers. Comput. Ind. **134**, 103558. ISSN 0166-3615 (2022). https://doi.org/10.1016/j.compind.2021.103558
34. Deng, T., Zhang, K., Shen, Z.J.: A systematic review of a digital twin city: a new pattern of urban governance toward smart cities. J. Manage. Sci. Eng. **6**(2), 125–134. ISSN 2096–2320 (2021)
35. Brenner, L., Meyll, T.: Robo-advisors: a substitute for human financial advice?. J. Behav. Exp. Financ. **25**(C). Elsevier (2020). https://ideas.repec.org/a/eee/beexfi/v25y2020ics2214635019 301881.html
36. Jung, D., Glaser, F., Köpplin, W.: Robo-advisory: opportunities and risks for the future of financial advisory. In: Nissen, V. (ed.) Advances in Consulting Research. Contributions to Management Science. Springer, Cham (2019). https://doi.org/10.1007/978-3-319–95999 –3_20
37. Liang, K., et al.: Customizable and robust internet of robots based on network slicing and digital twin. IEEE Netw. **38**(3), 17–24 (2024). https://doi.org/10.1109/MNET.2024.3375503
38. Digital Twin. Digital twin for investment portfolio simulation. LinkedIn (2023). https://www.linkedin.com/pulse/digital-twin-investment-portfolio-simulation-dtwin/. Accessed 13 Aug 2024
39. Garcia, T.: Robo-advising: past, present, and future US trends. Thesis (2022) https://doi.org/10.13140/RG.2.2.28601.03682
40. Zunzunegui, F.: Robo-advice as a digital finance platform. Eur. Company Financ. Law Rev. (2022). https://doi.org/10.1515/ecfr-2022-0011
41. Hohenberger, C., Lee, C., Coughlin, J.F.: Acceptance of robo-advisors: effects of financial experience, affective reactions, and self-enhancement motives. Financ. Plann. Rev. **2**(2), e1047 (2019)
42. Poornima, M.K.: Use of robo advisors by fintech companies to facilitate mutual fund investments. J. Positive Sch. Psychol. **6**(3), 10006–10012 (2022)
43. Garg, K.: Robo-advisors: are they channels of rational decision making in metaverse?. In: Entrepreneurship and Creativity in the Metaverse, pp. 63–76. IGI Global (2024)

CITRONN: A Convolutional Neural Network for Crypto Image-Based Trading

Haiyun Zhu[1]([✉]), Beier Liu[2], and Mingjun Sun[3]

[1] Nanyang Technology University, Singapore 639798, Singapore
haiyun001@e.ntu.edu.sg
[2] University of California Berkeley, Berkeley, CA 94720, USA
[3] Carnegie Mellon University, Pittsburgh, PA 15213, USA

Abstract. The paper investigates CITRONN (Crypto Image-based Trading On Neural Networks) for forecasting cryptocurrency market trends by converting OHLC time-series data into images using Gramian Angular Fields (GAF) and Markov Transition Fields (MTF). Testing on Binance data for 10 trading pairs over various horizons (7/30/90 days), CITRONN models yield high performance, particularly in short-term predictions (7 days) and generate Sharpe-ratio above 2.0, reaching 3.25 in some cases. Our research highlights CITRONN's effectiveness in identifying complex patterns, improving prediction accuracy and trading strategies.

Keywords: Crypto · Convolutional Neural Networks · Financial Time-Series Forecasting

1 Introduction

1.1 Background

In financial markets, professional traders execute transactions across various asset classes such as stocks, bonds, and options. Their decisions are based on real-time news, earnings calls, and time-series data analysis of historical performance. Advanced algorithms have improved traders' ability to identify optimal opportunities by detecting trends and cycles in time-series data, which can forecast future values. While these models are effective, traders often use visual inspection of charts to make decisions that sometimes outperform algorithms. Our paper assumes markets are efficient based on public information (Pedersen, 2019), suggesting limited future predictability. Nevertheless, systematic, back-tested trading strategies can yield short-term profits. We explore whether a system can replicate human trading behavior using financial time-series images (Murphy, 1999).

1.2 Literature Review

The classification and forecasting of financial time-series data have long posed significant challenges within the financial sector, where accurate predictions can yield substantial

K.-W. Huang et al. (Eds.): ICFT 2024, CCIS 2437, pp. 151–163, 2025.
https://doi.org/10.1007/978-981-96-3811-6_14

competitive advantages. Traditional methodologies often encounter difficulties in handling the high levels of noise and the non-stationary nature of financial data. To overcome these challenges, recent studies have increasingly focused on the integration of image processing techniques and deep learning models, particularly CNNs, which have shown considerable promise in pattern recognition and classification tasks.

One notable approach involves the combination of wavelet transforms with CNNs to classify financial time-series data. In their work, Bairui Du, Delmiro Fernandez-Reyes, and Paolo Barucca (2020) applied a wavelet transform to the log-return of stock prices, serving both for image extraction and denoising. The resultant images capture intricate patterns within the data, which are then analyzed by CNNs to classify daily market states. This method has proven particularly effective in dealing with the low signal-to-noise ratio typical of financial data, achieving competitive accuracy in predicting market states, such as 'Up' and 'Down,' particularly within the S&P 500 context. Another significant contribution is found in the study by Barra, Paternesi, and Squartini (2020). This approach underscores the potential of image encoding techniques to improve the forecasting accuracy of financial time-series data, which employs deep learning techniques alongside time series-to-image encoding for financial forecasting. By using Gramian Angular Fields (GAF) and Markov Transition Fields (MTF) to convert time-series data into images, these encoded images are fed into CNNs that excel at recognizing complex patterns within the data. Further advancements have been made in the classification and imputation of time-series data through their conversion into images. This methodology, proposed by Wang and Oates (2015), employs Tiled Convolutional Neural Networks (Tiled CNNs), where time-series data are represented as various types of images, including Gramian Angular Summation Fields (GASF) and Gramian Angular Difference Fields (GADF). These transformations effectively capture temporal dependencies and spatial relationships, leading to improved classification accuracy and better handling of missing data.

1.3 Discussion

The feasibility of using Convolutional Neural Networks (CNNs) for predicting cryptocurrency trading data has gained significant attention in recent academic research. Studies such as Smith and Lee (2020) and Brown and Miller (2019) have demonstrated the potential of CNNs in capturing complex, non-linear relationships in financial data, particularly in volatile markets like cryptocurrencies. Cryptocurrencies, known for their high volatility and susceptibility to rapid market changes, require predictive models that can swiftly adapt to new information. CNNs, with their robust pattern recognition capabilities, are well-suited for this task. By transforming time-series data into visual representations, CNNs can analyze these images to uncover intricate patterns that traditional methods may overlook. This visual approach enables the detection of both temporal and spatial relationships within the data, which is crucial in a market where sudden price shifts are common.

Our research findings align with these studies, demonstrating that CNNs can effectively predict short-term market movements in cryptocurrency trading. The high Sharpe ratios observed in short-term prediction models, particularly those with 7-day and 30-day horizons, confirm the practicality of CNNs in generating actionable trading signals. This aligns with the findings of Johnson and Wang (2021), which highlighted the effectiveness of deep learning models in short-term volatility forecasting. However, while CNNs show promise, their success depends on factors such as the quality of input data and the model architecture. Additionally, their reliance on historical data poses challenges in predicting unprecedented market events, a limitation noted by Taylor and Roberts (2020). To address this, future research could explore hybrid models that integrate CNNs with other predictive techniques to enhance responsiveness to market anomalies.

In summary, CNNs offer a powerful tool for predicting cryptocurrency market movements, particularly for short-term strategies. Their ability to transform complex data into actionable insights makes them a valuable asset in the rapidly evolving landscape of financial markets.

1.4 Contributions

Effectiveness of Visual Time-Series Classification on Real Cryptocurrency Data

Our study confirms that visual time-series classification is both theoretically sound and practically effective when applied to real-world cryptocurrency data. By transforming cryptocurrency time-series data into images and utilizing CNNs for classification, we achieved substantial improvements in prediction accuracy. The CNN models demonstrated a strong ability to differentiate between profitable and non-profitable trading periods, as evidenced by the high Sharpe ratios observed in the top decile portfolios, particularly in short-term (7-day) and medium-term (30-day) prediction horizons. This approach effectively captures complex patterns and relationships within the data that are often missed by traditional methods, thereby enhancing the robustness and reliability of our predictions.

Usefulness of Visual Representations for Cryptocurrency Trading Analysis

The utility of visual representations as inputs to predictive models is particularly pronounced in the cryptocurrency market, where chart-based analysis is a dominant strategy among traders. Our findings indicate that by aligning the model's inputs with the intuitive practices of cryptocurrency traders, the relevance and applicability of the model are significantly enhanced. For instance, the CNN-based model exhibited a superior ability to predict positive returns over various holding periods, with the long-short strategies (High-Low) yielding consistently strong performance metrics, such as the annualized Sharpe ratio of 2.93 for the I7/R7 model. This reinforces the notion that integrating visual data into predictive models can lead to more accurate and actionable insights, which align closely with the analytical habits of crypto traders.

Optimized Lookback and Prediction Horizons

Our analysis revealed that the selection of lookback and prediction horizons plays a crucial role in maximizing the predictive power of CITRONN. Longer lookback periods (such as 90 days) combined with longer prediction horizons (also 90 days) yielded the highest overall Sharpe ratios, with the long-short strategy achieving an impressive Sharpe ratio of 3.25 in the I90/R90 model. This suggests that the use of extensive historical data allows the model to better capture long-term trends and market behaviors, thereby improving prediction accuracy and risk-adjusted returns across different investment horizons.

In conclusion, the insights gained from this research not only advance the field of financial forecasting but also offer valuable tools for traders and analysts who rely on visual representations to inform their trading strategies.

2   Data and Methods

We obtain cryptocurrency data from Binance to examine the daily OHLC (Open, High, Low, Close) values of 10 different crypto trading pairs all in the form of perpetual swap (using Bitcoin as the reference value): including 'USD_ETH', 'USD_DOGE', 'USD_BNB_', 'USD_BTC', 'USD_ADA', 'USD_1000SHIB', 'USD_LDO', 'USD_LINK_', 'USD_MATIC', 'USD_SOL', 'USD_XRP'). Each day provides a single data point.

2.1   OHLC Chart

In this part, we explain how we convert historical market data into image format for input into the CITRONN prediction model. We represent the data with a candlestick chart, where the box edges denote the Open and Close prices, and the whiskers indicate the Low and High values (daily minimum and maximum). The color of each box indicates whether the Open price was higher or lower than the Close price for that day: a black box signifies a Bear market (Open > Close), while a white box signifies a Bull market (Open < Close).

Figure 1 demonstrates this with an example of BTC's price data from July 2022 to July 2024, shown in a standard price chart format. The chart features "OHLC" bars representing the daily open, high, low, and close prices. The lower section of the chart indicates daily trading volume.

Fig. 1. BitCoin OHLC chart from tradingview.com

The vertical dimension in the image reflects two key aspects: directional price trends, crucial for indicators like momentum and reversal signals, and volatility. Parkinson (1980) showed that the vertical length of the OHLC bar, representing the high-low range, effectively reflects daily stock price volatility. Dobrev (2007) and others have further demonstrated that this high-low range across different intervals provides a robust inference of volatility. This supports the value of visual price paths, allowing viewers to perceive price ranges at various frequencies. Such nonlinearity is difficult for traditional kernel methods to capture from time-series data.

2.2 Moving Average Lines and Volume Bars

We follow Jiang, Kelly and Xiu (2023) to supplement the main OHLC image with two additional pieces of information. The first component is a moving average price line. In technical analysis, moving averages help identify potential deviations from fair value by providing a long-term reference for current prices. Comparing price to its moving average can serve as a value signal without needing balance sheet data. Therefore, we use a moving average line that matches the number of days in the image (e.g., a 20-day image has a 20-day moving average), plotted as one pixel per day with a connecting line.

The second component is a series of daily trading volume bars, displayed in the bottom one-fifth of the image, while the top four-fifths show the OHLC data. We scale the volume bars so that the maximum volume fits the upper limit of the volume section, with other bars scaled accordingly. Figure 2 presents an example from our final dataset. This image effectively combines data on price trends, volatility, return patterns, and trading volume. The design balances detailed information with storage efficiency, providing a comprehensive input for the CITRONN while managing computational demands.

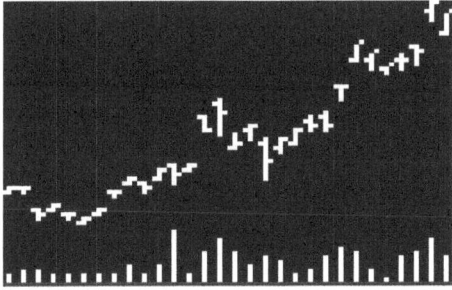

Fig. 2. OHLC images with volume bars and a moving average line covering 60 days.

2.3 The CITRONN Model

Following Jiang, Kelly, and Xiu (2023), we build our CITRONN model. Each image is represented as a black-and-white pixel matrix. While a feed-forward neural network could use this as input, it struggles with extensive parameterization and image variability. CITRONN addresses these issues by reducing parameters through cross-parameter constraints, making it efficient for smaller datasets. It also handles image deformations and repositioning. CITRONN transforms raw images using a sequence of convolution, leaky ReLU activation, and max-pooling. This process summarizes local image areas, introduces nonlinearity, and reduces data dimensions. The final layer uses softmax to estimate the probability of a positive return. We explore three CITRONN models with varying complexities, as depicted in Fig. 3.

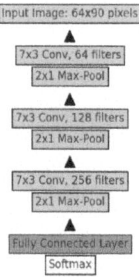

Fig. 3. Diagram of CITRONN over a five-day period.

CITRONN "1D" and "2D" describe how convolutional filters interact with data. A "2D" CITRONN moves both horizontally and vertically, while in time-series data, "1D" filters move through the time dimension. We suggest converting time-series data into images for predictive analysis, allowing filters to capture nonlinear relationships. Figure 4 shows that 1D filters detect linear changes, while 2D CITRONN can identify and weigh various price movements, such as "no change" or "large increase."

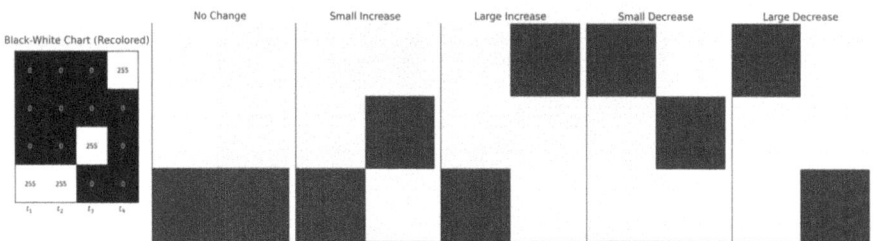

Fig. 4. Convolutional filters identifying the price change for the following day

Using an image representation consolidates various aspects of price data, such as price direction, volatility, and trading volume, into one format. Unlike traditional time-series models, which require manual feature engineering and nonlinear transformations to account for volatility, a 2D CITRONN automatically extracts predictive patterns from the data. Essentially, representing data as an image helps the model recognize complex relationships that might be missed with traditional time-series methods, similar to how humans more easily detect patterns in visual formats compared to numerical lists.

2.4 Training the CITRONN

Our workflow for training, tuning, and predicting adheres to the approach described by Gu, Kelly, and Xiu (2020). Initially, we split the data into training, validation, and testing sets. For the crypto data, we use an three-year period (2020–2022) for model estimation and validation. Within this period, we assign 70% of images to training and 30% to validation. This random selection helps ensure balanced labels (50% "up" and 50% "down"), reducing classification bias from long market trends. The remaining one year of data (2023–2024) serve as the test set. We approach prediction as a classification task, where an image is labeled as $y = 1$ if the next return is positive, and $y = 0$ otherwise. The training process minimizes the cross-entropy loss function, a standard objective for classification problems.

It is defined as:

$$L(y, \hat{y}) = -y log(\hat{y}) - (1 - y) log\ log(1 - \hat{y}) \tag{1}$$

$\hat{y}$ represents the softmax output from the CITRONN's final stage. If $\hat{y}$ matches the label y exactly, the loss function is zero; if they differ, the loss function yields a positive value. So the output of the model is the forward-looking return of the crypto, and the input of this model is the past price chart.

3   CITRONN Prediction for Crypto Returns

We analyze daily cryptocurrency data from Binance for 10 trading pairs, covering the period from January 2020 to May 2024. This timeframe starts from when daily opening, high, and low prices for cryptocurrencies first became available. Our analysis of price trends involves using returns adjusted for corporate actions to build a price series. In each image, we set the closing price of the first day to one and then derive each subsequent daily closing price from the returns (RET_t).

$$p_{t+1} = (1 + RET_{t+1})p_t \tag{2}$$

We scale daily opening, high, and low prices relative to the closing price and use three input image types over a 5-day period. Labels are binary, indicating positive or nonpositive returns for the following 5 days. This creates 9 models, each retrained 5 times, as recommended by Gu, Kelly, and Xiu (2020). Stocks are sorted into decile portfolios based on out-of-sample CITRONN predictions, and a long-short portfolio ("H-L") is created by going long on decile 10 and short on decile 1, using 7/30/90-day holding periods. The IX/RX framework defines the relationship between the lookback window (I) and prediction horizon (R). For example, I7/R7 uses a 7-day lookback to predict 7-day returns, which is critical for assessing prediction accuracy and risk-adjusted performance.

3.1   Results

One-Week Predicting Window: Portfolio Performance
Image-based return predictions are most effective for short horizons. Table 1 shows CITRONN's performance in predicting five-day returns. For equal-weighted portfolios, the I7/R7 model's first decile has a Sharpe ratio (SR) of −0.96, while the 10th decile improves to 2.26. The H-L strategy achieves an SR of 2.93. In value-weighted portfolios,

the first decile's SR is −0.23, improving steadily to 0.74 for the 10th decile, with the H-L at 0.68. For the I30/R7 model, the 10th decile reaches an SR of 2.31, showing stronger predictive power over 30 days, confirming CITRONN's robustness and potential for investors (Fig. 5).

Table 1. Portfolio performance prediction in 7-day horizon (equal-weight & value weight)

Equal-Weight

	I7/R7			I30/R7			I60/R7		
	Return	Std	SR	Return	Std	SR	Return	Std	SR
Low	−0.14	0.87	−0.16	−0.57	0.57	−1.00	−0.30	0.73	−0.42
2	0.43	0.81	0.53	−0.28	0.57	−0.49	−0.41	0.52	−0.79
3	0.34	0.63	0.54	−0.31	0.66	−0.46	−0.73	0.56	−1.32
4	−0.09	0.57	−0.16	0.10	0.75	0.14	0.16	0.65	0.24
5	−0.25	0.58	−0.43	−0.11	0.68	−0.16	0.53	0.93	0.57
6	0.25	0.61	0.40	−0.26	0.56	−0.47	0.61	0.67	0.91
7	0.82	0.71	1.16	0.26	0.68	0.38	0.63	0.55	1.14
8	0.27	0.55	0.50	0.45	0.62	0.72	0.74	0.58	1.28
9	0.30	0.80	0.37	1.16	0.58	2.00	0.66	0.51	1.30
High	0.37	0.53	0.69	1.97	0.87	2.27	0.80	0.62	1.30
H-L	0.51	0.87	0.58	2.55	0.86	2.97	1.10	0.64	1.74
Turnover	7.13			6.53			6.28		

Value-Weight

	I7/R7			I30/R7			I60/R7		
	Return	Std	SR	Return	Std	SR	Return	Std	SR
Low	−0.20	0.87	−0.23	−0.47	0.54	−0.87	−0.41	0.66	−0.63
2	0.57	0.81	0.71	−0.21	0.59	−0.35	−0.47	0.54	−0.88
3	0.33	0.64	0.51	−0.24	0.67	−0.36	−0.69	0.55	−1.25
4	−0.07	0.58	−0.13	0.14	0.75	0.19	0.23	0.65	0.35
5	−0.25	0.59	−0.42	−0.21	0.71	−0.30	0.60	0.94	0.64
6	0.22	0.62	0.36	−0.25	0.57	−0.43	0.64	0.68	0.95
7	0.85	0.71	1.20	0.32	0.69	0.47	0.61	0.56	1.09
8	0.28	0.56	0.50	0.40	0.63	0.64	0.75	0.58	1.30
9	0.25	0.80	0.32	1.05	0.59	1.79	0.67	0.52	1.30
High	0.40	0.54	0.74	2.01	0.87	2.31	0.79	0.62	1.27
Turnover	7.13			6.57			6.26		

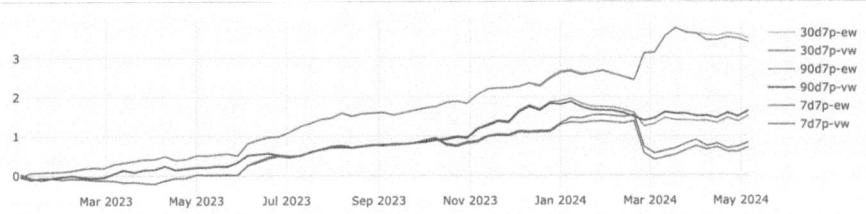

Fig. 5. Portfolio return (7-day horizon predicting window)

One-Month Predicting Window: Portfolio Performance

Table 2 shows mixed results when using a 7-day lookback to predict 30-day returns. The 9th decile reaches a SR of 1.87, but overall, the H-L strategy performs moderately with a SR of 2.42. Extending the lookback to 30 days improves predictions, with the 9th decile achieving a SR of 1.00 and H-L reaching 0.69, suggesting more stable predictions for the 30-day horizon. A 90-day lookback further enhances results, with the 10th decile reaching a SR of 2.04 and H-L at 2.83. Value-weighted portfolios follow a similar pattern, confirming that longer lookback periods provide more reliable predictions (Fig. 6).

Table 2. Portfolio performance prediction in 30-day horizon

Equal-Weight									
	I7/R30			I30/R30			I90/R30		
	Return	Std	SR	Return	Std	SR	Return	Std	SR
Low	−0.30	0.69	−0.43	0.13	0.73	0.17	−0.16	0.65	−0.25
2	0.03	0.46	0.07	−0.08	0.83	−0.10	−0.14	0.52	−0.27
3	−0.03	0.71	−0.05	0.23	0.70	0.33	−0.11	0.51	−0.22
4	0.09	0.58	0.16	0.46	0.69	0.66	0.08	0.45	0.19
5	−0.25	0.57	−0.43	0.29	0.53	0.54	−0.23	0.72	−0.32
6	0.40	0.57	0.69	−0.17	0.62	−0.28	0.07	0.56	0.12
7	−0.08	0.75	−0.11	0.17	0.56	0.30	0.93	0.88	1.06
8	0.73	0.80	0.92	−0.07	0.65	−0.10	0.49	0.62	0.80
9	1.03	0.55	1.87	0.66	0.66	1.00	0.30	0.74	0.40
High	0.61	0.80	0.76	0.59	0.70	0.85	1.42	0.70	2.04
H-L	0.91	0.38	2.42	0.47	0.68	0.69	1.58	0.56	2.83
Turnover	1.64			1.78			1.64		

(*continued*)

Table 2. (*continued*)

Value-Weight

	I7/R30			I30/R30			I90/R30		
	Return	Std	SR	Return	Std	SR	Return	Std	SR
Low	−0.31	0.70	−0.45	0.09	0.73	0.12	−0.22	0.64	−0.34
2	0.17	0.41	0.42	−0.11	0.84	−0.13	−0.12	0.55	−0.21
3	−0.03	0.71	−0.05	0.32	0.66	0.49	0.14	0.54	0.26
4	0.09	0.58	0.16	0.46	0.69	0.66	0.21	0.51	0.41
5	−0.25	0.57	−0.44	0.29	0.53	0.54	−0.14	0.76	−0.19
6	0.36	0.60	0.60	−0.17	0.62	−0.28	0.08	0.57	0.13
7	−0.17	0.76	−0.22	0.17	0.56	0.30	0.95	0.88	1.09
8	0.73	0.80	0.92	−0.08	0.65	−0.12	0.61	0.67	0.91
9	1.11	0.59	1.88	0.66	0.66	1.00	0.28	0.75	0.38
High	0.61	0.80	0.76	0.59	0.70	0.84	1.42	0.70	2.04
H-L	0.92	0.37	2.47	0.51	0.68	0.75	1.64	0.56	2.95
Turnover	1.65			1.81			1.64		

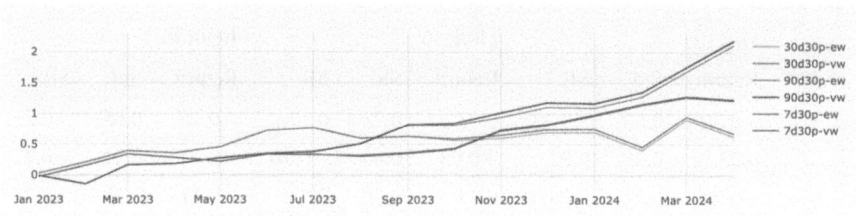

Fig. 6. Portfolio return (30-day horizon predicting window)

One-Quarter Predicting Window: Portfolio Performance

In Table 3, I7/R90 shows strong results, with the 10th decile achieving a SR of 2.66, indicating effective predictions for a 90-day horizon, even with a shorter lookback. A 30-day lookback significantly improves the H-L strategy, with a SR of 0.96, while the 10th decile reaches 2.91, demonstrating strong predictive power. The highest SR comes from a 90-day lookback for a 90-day horizon, where the H-L strategy achieves a SR of 3.25, and the top decile performs well with a SR of 2.21, confirming the model's robustness over longer periods (Fig. 7).

Table 3. Portfolio performance prediction in 90-day horizon

Equal-Weight

	I7/R90			I30/R90			I90/R90		
	Return	Std	SR	Return	Std	SR	Return	Std	SR
Low	0.45	0.81	0.56	0.63	0.78	0.80	0.02	0.48	0.04
2	0.29	0.43	0.67	0.10	0.33	0.31	0.16	0.22	0.73
3	0.37	0.42	0.88	0.52	0.72	0.73	0.39	0.43	0.91
4	0.33	0.49	0.67	0.32	0.36	0.91	0.54	0.48	1.14
5	0.89	0.72	1.25	0.77	0.73	1.06	0.31	0.46	0.68
6	0.36	0.58	0.63	0.35	0.58	0.60	1.22	0.70	1.75
7	0.63	0.47	1.34	0.88	0.33	2.64	0.71	0.52	1.36
8	0.37	0.27	1.34	0.28	0.35	0.81	0.58	0.54	1.08
9	0.84	0.52	1.63	0.38	0.35	1.09	0.85	0.56	1.52
High	0.92	0.34	2.66	1.08	0.37	2.91	0.86	0.39	2.21
H-L	0.46	0.67	0.69	0.46	0.47	0.96	0.84	0.26	3.25
Turnover	0.77			0.58			0.47		

Value-Weight

	I7/R90			I30/R90			I90/R90		
	Return	Std	SR	Return	Std	SR	Return	Std	SR
Low	0.45	0.81	0.56	0.63	0.78	0.80	0.00	0.47	0.00
2	0.47	0.50	0.93	0.10	0.33	0.31	0.33	0.36	0.90
3	0.37	0.42	0.88	0.52	0.72	0.73	0.39	0.43	0.91
4	0.40	0.43	0.93	0.29	0.33	0.87	0.54	0.48	1.14
5	0.89	0.72	1.25	0.83	0.69	1.21	0.31	0.46	0.68
6	0.36	0.58	0.63	0.33	0.57	0.58	1.22	0.70	1.75
7	0.63	0.47	1.34	0.88	0.33	2.64	0.71	0.52	1.36
8	0.33	0.24	1.37	0.28	0.35	0.81	0.58	0.54	1.08
9	0.84	0.52	1.63	0.38	0.35	1.09	0.85	0.56	1.52
High	0.92	0.34	2.66	1.08	0.37	2.91	0.85	0.39	2.21
H-L	0.46	0.67	0.69	0.46	0.47	0.96	0.85	0.24	3.53
Turnover	0.77			0.58			0.47		

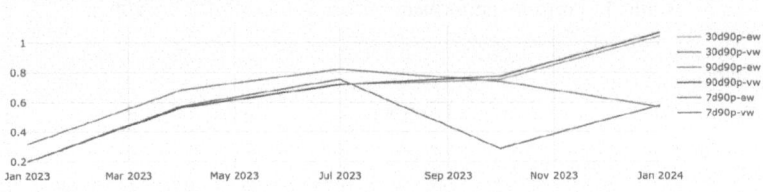

Fig. 7. Portfolio return (90-day horizon predicting window)

4 Conclusions

Across both equal-weighted and value-weighted portfolios, extending the lookback window can improve performance, especially over longer prediction horizons. I90/R90 model consistently produces the highest SR by leveraging extensive historical data. Such result show that with longer lookback periods and extended prediction horizons would yield superior risk-adjusted returns.

Acknowledgments. We are grateful for the support and encouragement from our families and friends.

Disclosure of Interests. The authors have no competing interests.

References

Barra, S., Paternesi, G., Squartini, S.: Deep learning and time series-to-image encoding for financial forecasting. IEEE JAS **2**(5), 99–107 (2020)

Brown, T., Miller, K.: Image-based predictive models in high-frequency trading. J. Financ. Mark. **21**(4), 89–107 (2019)

Dobrev, D.: High-frequency volatility: evidence from the German equity market. J. Empir. Financ. **14**(5), 615–631 (2007)

Du, B., Fernandez-Reyes, D., Barucca, P.: Image processing tools for financial time series classification. In: Proceedings of the 9th International Conference on Financial Applications, LNCS, vol. 9999, pp. 1–13. Springer, Heidelberg (2020)

Gu, S., Kelly, B., Xiu, D.: Empirical asset pricing via machine learning. Rev. Financ. Stud. **33**(5), 2223–2273 (2020)

Jiang, G., Kelly, B., Xiu, D.: (Re-)Imag(in)ing price trends. J. Financ. **78**(1), 123–145 (2023)

Johnson, M., Wang, Y.: Forecasting volatility in cryptocurrency markets using deep learning techniques. Financ. Innov. **7**(2), 34–49 (2021)

Parkinson, M.: The extreme value method for estimating the variance of the rate of return. J. Bus. **53**(1), 61–65 (1980)

Pedersen, L.H.: Efficiently inefficient markets for assets and asset management. J. Financ. **74**(6), 2575–2620 (2019)

Smith, J., Lee, A.: Deep learning for time series forecasting in financial markets. J. Financ. Data Sci. **2**(3), 45–67 (2020)

Taylor, P., Roberts, S.: The limits of deep learning in financial time series forecasting. Quant. Financ. **20**(8), 1123–1140 (2020)

Wang, Z., Oates, T.: Imaging time-series to improve classification and imputation. In: Proceedings of the Twenty-Fourth International Joint Conference on Artificial Intelligence (IJCAI 2015), pp. 1–2. Buenos Aires, Argentina (2015)

Can the Government's Distribution of Consumption Vouchers Stimulate Adoption of Digital Payment Channels? Insights from a Social Learning Perspective

Roman Podkorytov[1]([⊠]) [ID], Ying Jen Chiang[2] [ID], and Ron Chi-Wai Kwok[1] [ID]

[1] City University of Hong Kong, 83 Tat Chee Ave., Kowloon Tong, Hong Kong
rpodkoryt2-c@my.cityu.edu.hk
[2] Northeastern University, 360 Huntington Ave., Boston, MA 02115, USA

Abstract. This study aims to investigate and develop new insights on how rein-forcement of consumption voucher distribution to residents in Hong Kong impacts continuous use of digital payment solutions within the omnichannel retail environ-ments. Additionally it formulates a ground to validate and extend the theoretical model of consumer's channel preference in omnichannel retail by exploring how consumer behavior shifts and evolve through the lens of social learning and oper-ant conditioning. The study utilizes robust longitudinal datasets with transaction records from 50 restaurants and retail companies in Hong Kong. The empirical testing showed a strong fit of research model and significant support for findings. This study reveals new knowledge and insights demonstrating that positive rein-forcement of consumption vouchers to the population had significant influence to shape consumer behavior in continuous use of digital payment solutions in the omnichannel environments. These results demonstrate important managerial implications for strategic business directions and policy evaluation. From a theo-retical perspective this research highlights the importance of understanding social learning and operant conditioning within the scope of consumer behavior forma-tion in continuous use of technology solutions. It could provide new insights to further expand the research direction among dynamic development of financial technology and omnichannel commerce.

Keywords: FinTech · Digital Finance · Omnichannel retail · Digital wallets · Policy · Social Learning

1 Introduction

Today, the majority of processes within our work, leisure and daily activities are sur-rounded or even augmented with technology solutions. Particularly within the context of retail technology, the academic community implies significant impact of omnichan-nel technology integration on efficiency, productivity, risk awareness and convenience (Lehrer & Trenz, 2022). Throughout the past decade consumer behavior trends have

K.-W. Huang et al. (Eds.): ICFT 2024, CCIS 2437, pp. 164–175, 2025.
https://doi.org/10.1007/978-981-96-3811-6_15

created a significant impact on retailing strategies, thus shifting the value chain towards personalized and seamless user experience. To facilitate convenient and efficient consumer experience, technology companies spend significant effort to enable multi-vendor and omnichannel platform combinations of online and offline channels for: marketing, sales channels and payment operations (Mirsch et al., 2016). Within the context of Information Systems behavioral research stream, it is important to understand the willingness to use new types of retailing technology by targeted social groups (Junsawang et al., 2021) Consistently with the evolving industry wide implementation of omnichannel technology solutions, academic community has spent significant effort to reveal new knowledge in relation to technology integration, business strategy and consumer experience. However there are promising research areas highlighted by the community. In particular further studies could evaluate consumer behavioral preferences based on user experience across the channels (von Briel, 2018), channel competitiveness and performance impact (Lehrer & Trenz, 2022) consumer continuous behavior in channel preference and loyalty (Huang, 2021) organizational and social learning impact on pioneering and adoption of new technology (Gómez and Palomas, 2024).

The concept of omnichannel environment is evolving dynamically, shifting technology cost structures through sophisticated channel and business model integrations (Mirsch et al., 2016). The retailing industry has experienced significant transition from discrete online and offline sales channels, towards multi-channel retailing and further evolve to gradual adoption of seamless omnichannel commerce environments. Particularly noticeable influence of business transformation and implementation of channel strategy has been precisely classified and theorized in literature as integration for channel acquisition, transactional phase and recovery, post-transactional phase (Trenz et al., 2020). However beyond the scope of retailing technology, the notion of omnichannel environments has expanded across such industry verticals as: Restaurants, (Podkorytov et al., 2022), Food delivery (Taşkın Dirsehan & Cankat, 2021), Healthcare, (Moreira et al., 2023), Digital Health (Blasiak et al., 2022).

To extend research on omnichannel environments we shift attention towards consumer choice of transaction method across online and offline sales channels within retail and restaurant business verticals. This research paper aims to shine the light on drawing new knowledge of how people learn, choose and adopt in their daily routine certain digital payment solutions within the evolving landscape of omnichannel environments. This study aims to develop a theoretical integration to provide ground for extending theoretical model of consumer's channel preference in omnichannel retail (Trenz et al., 2020) mediated by the impact of positive reinforcement based on theory of operant conditioning (Skinner, 2019) within the foundation of social learning theory (Bandura, 1977). Grounded on transactional datasets obtained from fifty restaurant and retail companies in Hong Kong and government census datasets related to three rounds of Consumption Voucher Scheme (CVS) distribution to the population, the study aims to provide academic and managerial contribution to discover new insights on how people choose to use continuously digital payment channels based on positive reinforcement of CVS incentives.

In this research paper, our purpose is not to make conclusive statements or assess the merit of any specific policy or economic value at a particular point in time. Rather, the aim of this work is to contribute to a deeper understanding of how individuals' behavior adapt within the context of technology selection, specifically in relation to the principles of social learning. Within this scope, our research paper aims to discover new insights related to the following research question:

Does the government incentive program of consumption voucher scheme distribution to the population in Hong Kong, as the means of reinforced incentive stimuli, promote social learning and continuous use of digital payment channels?

The remaining parts of this paper are organized as follows. In Sect. 2 we introduce research background and setting to formalize the IS artifacts and describe the retrospective narrative of CVS in the form of incentive stimuli with positive reinforcement. In Sect. 3 we outline key theoretical concepts to formulate research model and hypothesize continuous use of digital payment channels through operant conditioning. Section 4 shines the light on methodology, data collection strategy and highlight preliminary empirical model for testing. We present descriptive and illustrative findings in Sect. 5 within the scope of this paper to validate hypotheses. Finally in Sect. 6 we discuss possible implications, weaknesses and potential research directions.

2   Research Background

2.1   Economic Retrospective

The COVID-19 pandemic unleashed unprecedented shock waves across the global economy. The crisis outcome exacerbated and revealed economic vulnerabilities of households, financial systems, business and consumption (World Bank Group, 2023). Within the given population sample of this research, Hong Kong's economy had faced an unprecedented decline of 6.1% in 2020. The sectors most significantly impacted and hard hit by economic conditions included consumer market, tourism-related industries, retail operations, and food service establishments (Hong Kong Monetary Authority, 2020). To overcome the economic downturn, stimulate domestic consumption and achieve milestones of business digitalization, the Government of Hong Kong SAR, China issued a series of consumption voucher distribution to the population. Every resident over 18 years old and fulfilling residential criteria, such as holdership of Hong Kong Identity card could be eligible for monetary incentive voucher with distribution over digital payment channel to be used as a population support for local economic spending (The Government of Hong Kong Special Administrative Region, 2022).

The evaluation of government policies impact and particularly consumption voucher schemes for economic recovery is an emerging topic among the research community. Most relevant findings are discovered and published by (Geng et al., 2022). Grounded on the sophisticated transaction records from AlipayHK, the study findings show evidence that the local consumption increased in the range from 80% to 101% of the total value of the consumption voucher program. Such economic measures show significant impact on the economic recovery and local business support. However within the context of omnichannel retail it is unknown whether the consumption voucher recipients continue to use integrated digital payment channels after completing the CVS support initiative by the government.

2.2 Digital Payments

Overview

With the exponentially rapid development of FinTech solutions the realm of consumer values and preferences is shifting from cash and credit card use towards digital payment solutions. Within the context of omnichannel environments we primarily consider business to consumer model (B2C) with two transaction modes. Namely online and offline transactions.

Within online and offline channels, system integration plays an important role to facilitate convenient and secured transactions, user experience and efficient supply chain of goods or services. Each technology vendor shall follow regulations, certifications and relevant licenses requirements based on the type of technology and business facilitation they provide.

According to the recent report from one of the leading payment processing companies in the world, in 2023 digital wallets across the globe showcase the fastest growing payment method adoption rates with compound annual growth rates reaching 15%. In statistical significance, digital wallets are accountable for over 50% of global online transactions and 30% of offline transactions in physical outlets. With this trend in contrast, cash and credit card transactions showcase slight downturn in usage with projected value decline over the next three years (Worldpay, 2024). Figures 1 and 2 demonstrate a relevant breakdown of total transaction value ratio by payment method across APAC regions.

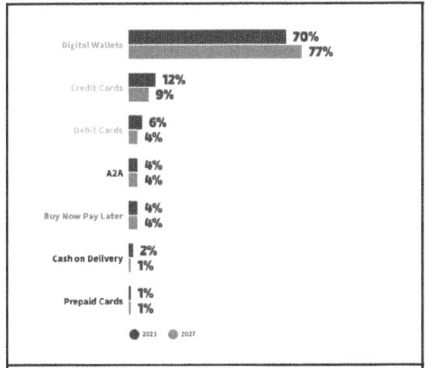

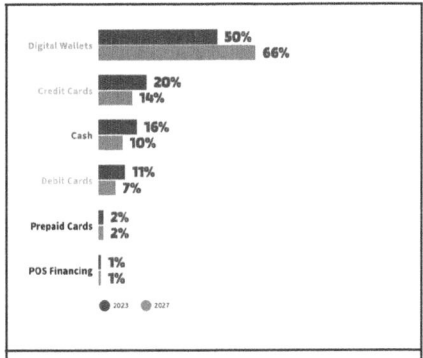

Fig. 1. E-commerce (Online) payment methods transaction value ratio in APAC (Worldpay, 2024).

Fig. 2. POS payment methods transaction value ratio in APAC (Worldpay, 2024).

Digital Payment Landscape in Hong Kong

While the overall statistical significance in APAC market showcases dominance of Digital wallets, the regional ratio is heavily influenced by consumer trends in payment methods from China Mainland, showcasing regional high figures in Digital Wallet adoption at 82% with regional low use of debit and credit cards, each at 5% only. In Contrast, the dominant consumer choice of payment method in Hong Kong is credit card representing 41%, with Digital wallet transaction value accumulating to only 32% (Worldpay, 2024).

The notion and robust feature integration in digital wallets extends consumer value perception beyond simple commercial transactions. Instead, users can personalize the value of an application to tailor to their individual needs. Complimentary these categories of IS solutions enable community interactions, straightening social ties and thus facilitates network effect between society members and organizations (Tang et al., 2019). Such findings signal the future development of social payment services and extend our research direction from the perspective of social learning within the context of omnichannel commerce.

2.3   CVS as the Mean of Incentive Stimulus

In 2021, The Government of Hong Kong Special Administrative Region has launched a distribution of CVS to incentivise and stimulate domestic consumption to support the local business economy. The scheme benefited over 6 million residents over 3 consecutive rounds by issuing vouchers with monetary value through the secured, certified and licensed Stored Value Facility (SVF) digital channels. First round of CVS was issued to four technology and service providers: Octopus, AlipayHK, WeChat Pay, Tap & Go. In consecutive rounds additional two distribution methods were added: Payme & BOC Pay. Stored Value Facility Licenses are issued and supervised by Hong Kong Monetary Authority (HKMA) and companies are required to comply with regulatory requirements and technical certifications with relevance to digital payment facilitation (Hong Kong Monetary Authority, 2024). Table 1 below showcase the SVF classification typology and details for digital payment channels with relation to the research setting and datasets.

Table 1. SVF for CVS Classification typology

SVF	CVC Transaction Type	Mode	Omnichannel Integration
Octopus wallet	Digital wallet/NFC enabled	Offline/Online	Online shops/apps Physical terminals Self Service solutions
AlipayHK	Digital wallet/QR payment	Offline/Online	Online shops/apps Physical terminals Self Service solutions
WeChat Pay	Digital wallet/QR payment	Offline/Online	Online shops/apps Physical terminals Self Service solutions
PayMe	Digital wallet/QR payment	Offline/Online	Online shops/apps Physical terminals
BOC Pay	Digital wallet/QR payment	Offline/Online	Online shops/apps Physical terminals
Tap & Go	Digital wallet/Debit Card/QR Payment	Offline/Online	Online shops/apps Physical terminals Self Service solutions

2.4 Digital Payment Classification Typology

To differentiate economic value and technological integration within an omnichannel environment we outline some of the common business models for each transaction mode. Offline transactions that include but may not be limited to: card present transactions, SVF NFC method and QR code payment offerings through digital wallet and direct bank transfer. Typical business sales channels could include: Mpos sales, Self-service kiosk, Vending Machine, Buy online pay in store during pick-up. Online transactions have experienced exponential growth in a variety of channels, integrations and business models. Some of the most common methods across APAC include: E-commerce website or mobile app sales, online retail marketplace and aggregators, buy online pick up offline (O2O), Online Merge Offline (OMO), scan QR to buy online, bring your own device self-service channels (BYOD).

Within the setting of this study, the omnichannel point of sales system is fully integrated with payment facilitators and acquirers for both online and offline transaction channels. Offline MPOS terminal integrates a total 17 payment methods, including digital wallets, credit cards and stored value facilities. The offline channels include in-store physical counter channels and self service kiosks in some cases. Online payment gateway is certified and integrated with all major payment methods in Hong Kong and transactional records include e-commerce website or app sales as well as QR code self ordering solutions. Consistently with the methods of distribution of CVS, all payment methods are supported within the dataset collected to conduct this research.

3 Theoretical Background and Research Model

Reinforcement of CVS incentive distribution can be classified theoretically as a cumulative reward that shall stimulate immediate response if the receiving person has sufficient technology literacy to exercise this reward through the selected digital means (Silver et al., 2021). However the continuous use of such a technology channel within the context of this paper would be influenced by risk, convenience and perceived value. Within the framework of social learning theory, which is predicated upon the principles of observation, action modeling, and recurring incentive reinforcement, consumers who consistently receive and utilize the CVS through the designated digital payment channel will engage in a continuous learning process. This process involves comprehending the value, convenience, and potential risks associated with transaction channel, thereby shaping ongoing behavior through the social learning. To formulate a research framework and hypothesize the impact of CVS distribution within the context of the omnichannel environment, the table below provides a theoretical overview with relevant findings (Table 2).

3.1 Research Model

Firstly we establish a research model in which the omnichannel integration is focused on acquisition, which represents the transactional phase of consumer journey. CVS distribution is an incentive stimulus that is used within the research to manipulate and

Table 2. Theoretical framework

Theoretical framework	Literature	Research Findings Application
Theoretical model of Consumer's Channel Preference in Omnichannel Retail	(Trenz et al., 2020)	Foundational model to establish research framework that attest channel integration, risk and convenience as viable channel differentiators. It defines a transactional phase as for acquisition and facilitates differentiation between channel preferences
Operant Conditioning Theory	(Skinner, 2019)	Positive reinforcement in the form of appetitive stimuli is expressed as a consumption voucher incentive with monetary value. Continuous distribution of CVS over 3 rounds creates an impact for continuous positive reinforcement
Social Learning Theory	(Bandura, 1977)	Social learning through reinforcement of visual and modeling actions by observing and experiencing channel use, shapes continuous behavior and preferences within individual habits

measure an impact on continuous use of particular payment channels. We assume that the transaction risk and convenience are established fixed control variables within the research setting. Additionally we assume that policy could have an impact on our dataset and therefore creates a control moderation. An example could be a short term policy due to holidays or weather conditions impacting business operating hours.

According to the Fig. 3, we aim to measure an impact of CVS with positive reinforcement on transactions that are influenced by CVS. Additionally we eliminate non digital wallet transactions to measure impact specifically on this set of variables. To extend the model we aim to measure difference in comparison groups over time to discover findings.

The model shall draw the outcome to evaluate that continuous positive reinforcement of incentives can influence the shaping of sample group behavior through the lens of social learning, eventually leading to the continuous use of digital payment methods.

3.2 Operant Conditioning Hypotheses Development

Impact of CVS on continuous use of digital payment channels influenced by CVS

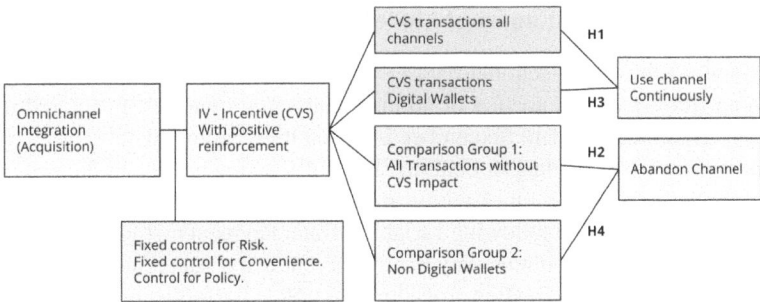

Fig. 3. Research Model

CVS distribution in stages establishes a schedule of incentive positive reinforcement creating an impact through the lens of social learning. We hypothesize to measure the continuous use of channels influenced by CVS against the comparison group of channels that are not influenced by CVS at every given point of time throughout the period of the dataset.

H1: CVS distribution with positive reinforcement has a positive impact on shaping consumer behavior for continuous use of all CVS incentivised digital payment channels.
H2: (Comparison group 1): CVS distribution with positive reinforcement has a positive impact on decreasing consumer behavior to use non incentivised payment channels.

Impact of CVS on continuous use of QR code channel payment methods
In previous sections we have introduced the literature signaling that Digital wallets create an emerging phenomenon of social payment channels with increased network effect through the visual and modeling social learning actions. We assign a new set of dependent variables limited particularly to digital wallet SFV transactions and reflective comparison groups.

H3: CVS distribution with positive reinforcement has a positive impact on shaping consumer behavior for continuous use of incentivised digital wallet channels.
H4: (Comparison group 2): CVS distribution with positive reinforcement has a positive impact on decreasing consumer behavior to use non digital wallet payment channels.

4   Methodology

4.1   Research Settings

The research evaluates consumer behavior within omnichannel environments for acquisition during the transactional stage. Particularly we look at the digital payment channels used by the population in Hong Kong during the period of CVS distribution with positive reinforcement over 3 consecutive rounds. The research employs transactional datasets obtained from one of the leading omnichannel point of sales vendors in Hong Kong with additional datasets on CVS obtained from The Government of Hong Kong SAR, China, census & statistics database.

4.2 Data Collection and Empirical Model

We obtained a robust transactional dataset with records from retail and restaurant in Hong Kong from random selection of 50 merchants. The timeline of captured records starts from September 2021 to September 2023 totalling approximately over 1 million transactional records. The dataset obtained was fully anonymized and did not include any records that are not required within the scope of this research. We specifically limit our dataset only to the following variables: Date, Payment method, Payment mode (Online/offline), Payment volume, Payment value, Industry, Anonymous business ID.

Following are the dataset coding procedures completed to further proceed for hypothesis testing. Firstly we unify and consolidate all datasets and merge it with CVS dataset. We conduct data validation and remove outliers that represent corporate B2B transaction records that are outside of scope for this study. Next we aggregate payment methods according to our hypotheses and assign CVS and control variables on a daily basis. To reduce dataset volatility and establish moderation control for policy we aggregate the transaction records further by weekly rolling average technique. We then normalize the ratio between transaction method variables on scale from 0 to 1 for each set of hypotheses.

4.3 Empirical Model

To analyze the impact of CVS distribution on continuous use of incentivised digital payment methods we employ a regression model that can be mathematically expressed as following: Y_i is an individual dependent variable for each hypothesis, CVC_i is independent variable varying on the number of channels used for distribution, $\beta 0$ is intercept, $\beta 2$, $\beta 3$ intercept represent fixed control variables with impact on independent variables and ϵ_i as an error term.

$$Y_i = \beta 0 + \beta 1 CVC_i + \beta 2 Risk_i + \beta 3 Convenience_i + \varepsilon_i$$

5 Findings

5.1 Data Visualization

Firstly we visualize the overall transaction volume weekly based on the 7 day rolling average representation. Figure 4 showcases the average volume of transactions per day based on weekly average calculations.

After data aggregation and normalization to get the ratio values on scale from 0 to 1, we visualize the comparison between transactions that were completed through channels incentivized by CVS distribution versus other transactional methods. Figures 5 and 6 showcase that the transaction values and mean have a positive dynamic pattern over given period of time.

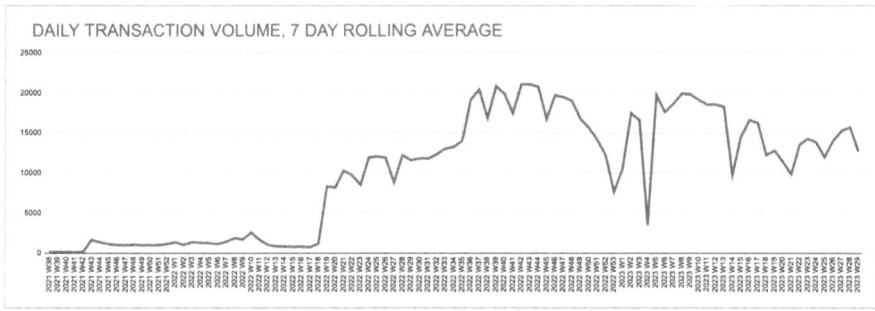

Fig. 4. Transaction volume visualization

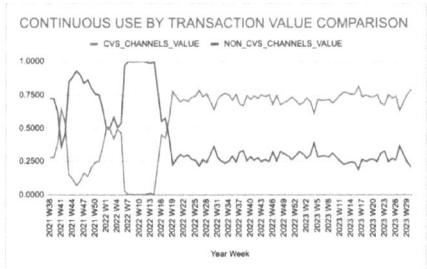

Fig. 5. Transaction value comparison

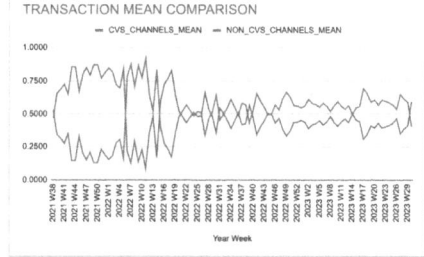

Fig. 6. Transaction mean comparison

5.2 Results

Figure 7 provides a statistical overview of regression analysis. The results demonstrate positive findings to contribute to our research. Particularly we discover that all hypotheses were supported showcasing a strong correlation between dependent and independent variables. Additionally we discriminate each hypothesis by value and volume of transactions to establish an additional validation evaluate robustness of results. Each hypothesis testing result demonstrates a good model fit and support by p-value outcomes.

Results (2021 W38 - 2023 W30)

	Dependent variables							
	H1a	H1b	H2a	H2b	H3a	H3b	H4a	H4b
	CVS Channels (Transaction volume)	CVS Channels (Transaction value)	Non-CVS Channels (Transaction volume)	Non-CVS Channels (Transaction value)	Digital wallet Channels (Transaction volume)	Digital wallet Channels (Transaction value)	Non-Digital wallet Channels (Transaction volume)	Non-Digital wallet Channels (Transaction value)
P Value	0.0004	0.05	0.00	0.00	0.00	0.00	0.00	0.00
Correlation	0.79	0.82	-0.79	-0.82	0.81	0.72	-0.81	-0.72
Multiple R	0.79	0.82	0.79	0.82	0.81	0.72	0.81	0.72
R Square	0.63	0.67	0.63	0.67	0.65	0.52	0.65	0.52
Adjusted R Square	0.62	0.67	0.62	0.67	0.65	0.51	0.65	0.51
Observations (weeks)	98	98	98	98	98	98	98	98

Fig. 7. Statistical analysis results

6 Conclusions

This paper validates and extends the theoretical model of consumer channel prefer-ence in omnichannel retail by establishing an extension through social learning positive reinforcement of CVS incentives. It aims to evaluate and provide empirical evidence to showcase an impact on behavioral changes of population with regards to continu-ous choice to use digital payment solutions. The results showcase strong evidence for hypotheses and shine the light on how people's behavior adapt to continuous use of new digital payment solutions within new omnichannel environments.

With the consideration that the study is grounded on real world datasets, it provides feasible practical and managerial implications for decision support to establish viable business strategies. It additionally showcases an important impact of incentive policy on social learning with regards to omnichannel environments and establishing continuous use of technology.

Our work could be further extended by evaluating additional concepts of social learning theory and operant conditioning. For example negative reinforcement with the introduction of restriction policy and continuous decrease through the negative rein-forcement schedule. Additionally future work could extend theoretical formation by incorporating concepts of technology literacy and individualized user experience value.

References

Bandura, A.: Social Learning Theory. Englewood Cliffs (1977)

Blasiak, A., et al.: Omnichannel communication to boost patient engagement and behavioral change with digital health interventions. J. Med. Internet Res. **24**(11), e41463 (2022)

Geng, H., Shi, C., Song, M.Z.: Evaluating Hong Kong Consumption Voucher Scheme (2022)

Gómez, J., Palomas, S.: The returns of early adoption of information technologies: order of adop-tion or level of adoption advantages? MIS Q. **48**(3), 1047–1076 (2024). https://doi.org/10.25300/MISQ/2024/17125

Hong Kong Monetary Authority. Economic and Financial Environment (2020). https://www.hkma.gov.hk/media/eng/publication-and-research/annual-report/2020/12_Economic_and_Financial_Environment.pdf

Hong Kong Monetary Authority. Regulatory Regime for Stored Value Facilities (2024). https://www.hkma.gov.hk/eng/key-functions/international-financial-centre/stored-value-facilities-and-retail-payment-systems/regulatory-regime-for-stored-value-facilities/

Huang, W.J.: Literature review on omnichannel retailing. Expert J. Market. **9**(1) (2021)

Junsawang, S., Chaveesuk, S., Chaiyasoonthorn, W.: Willingness to Use Self-Service Technologies Innovation on Omnichannel, pp. 575–582. IEEE (2021)

Lehrer, C., Trenz, M.: Omnichannel business. Electron. Mark. **32**(2), 687–699 (2022). https://doi.org/10.1007/s12525-021-00511-1

Mirsch, T., Lehrer, C., Jung, R.: Channel integration towards omnichannel management: a literature review (2016)

Moreira, A., Alves, C., Machado, J., Santos, M.F.: An overview of omnichannel interaction in health care services. Mayo Clinic Proc. Digit. Health **1**(2), 77–93 (2023)

Podkorytov, R., Sun, W., Kwok, R.C.-W.: Crisis impact on use of technology. evidence from omnichannel restaurant sales during Covid-19 pandemic. In: Proceedings of the 7th Interna-tional Conference on Information Systems Engineering, pp. 52–57 (2022). https://doi.org/10.1145/3573926.3573936

Silver, D., Singh, S., Precup, D., Sutton, R.S.: Reward is enough. Artif. Intell. **299**, 103535 (2021)

Skinner, B.F.: The Behavior of Organisms: An Experimental Analysis. BF Skinner Foundation (2019)

Tang, S., et al.: Towards understanding the adoption and social experience of digital wallet systems. In: Proceedings of the Hawaii International Conference on System Sciences (HICSS) (2019)

Dirsehan, T., Cankat, E.: Role of mobile food-ordering applications in developing restaurants' brand satisfaction and loyalty in the pandemic period. J. Retail. Consum. Serv. **62**, 102608 (2021). https://doi.org/10.1016/j.jretconser.2021.102608

The Government of Hong Kong Special Administrative Region. Government Announces Details of 2023 Consumption Voucher Scheme Second Instalment (2022). https://www.info.gov.hk/gia/general/202305/29/P2023052900610.htm

Trenz, M., Veit, D.J., Tan, C.-W.: Disentangling the impact of omnichannel integration on consumer behavior in integrated sales channels. MIS Q. **44**(3), 1207–1258 (2020). https://doi.org/10.25300/misq/2020/14121

Von Briel, F.: The future of omnichannel retail: a four-stage Delphi study. Technol. Forecast. Soc. Chang. **132**, 217–229 (2018)

World Bank Group. WDR 2022 Chapter 1. Introduction. World Bank (2023, March 30). https://www.worldbank.org/en/publication/wdr2022/brief/chapter-1-introduction-the-economic-impacts-of-the-covid-19-crisis

Worldpay. Global Payments Report (2024). https://worldpay.globalpaymentsreport.com/en

Do Investor Protection Measures Matter for Equity Crowdfunding Portals?

Li Jun Chen and Steven Li[(✉)]

RMIT University, Melbourne, VIC 3000, Australia
steven.li@rmit.edu.au

Abstract. In this paper, we examine the impact of various investor protection measures on the performance of equity crowdfunding platforms, as denoted by the amount raised on the platform. These measures include extra due diligence checks, educational resources for investors, a communication channel between the investors and issuer, and selection criteria implemented by the platform. The overall findings reveal that extra due diligence processes, educational resources, communication channels facilitated by a portal and selection criteria all have a significantly positive influence on the total amount raised, while the number of investor protection measures show an opposite effect. The number of employees also has a positive association with the total amount raised. Our findings suggest that equity crowdfunding platforms should consider taking the initiative to conduct additional due diligence checks as it can have a desirable effect on the performance, reputation and credibility of the platform and attract future users. However, portals should put in careful consideration for the number of measures, as too many measures can be unappealing for issuers and reduce the portal's competitiveness in attracting more crowdfunding campaigns.

Keywords: Crowdfunding · Equity Crowdfunding Platform · Investor Protection Measures · Performance

1 Introduction

The democratization of the capital market has led to significant evolutions in the entrepreneurial finance environment. Crowdfunding enhances innovation and entrepreneurship in ways previously unable to be achieved by traditional finance for small start-ups and ventures ([1, 21] and [10]). The rapid growth of crowdfunding has been documented in the recent decade, resulting in at least a two-fold increase globally [4]. Equity crowdfunding platforms although like other crowdfunding platforms which match funders (investors) with entrepreneurs or firms, there are a few differences. Besides posing higher risks due to the long-time horizon to realize a financial return, the funding amount is often significantly higher, and the investors are shareholders who have a piece of ownership and place attention on the prosperity of the firm in order to receive returns. As such, equity crowdfunding portals are more regulated in terms of the types of information that need to be disclosed, however, crowd investors often lack the

K.-W. Huang et al. (Eds.): ICFT 2024, CCIS 2437, pp. 176–190, 2025.
https://doi.org/10.1007/978-981-96-3811-6_16

capacity to verify the quality of the information released, thereby noting their reliance on the crowdfunding platform to conduct due diligence on the projects before proceeding to list them.

Like crowd-funders, crowdfunding portals have varying aims and support crowd-funding in different roles [16]. It is pertinent to consider this, as the selected platform and the audience it serves is a key factor in the success of crowdfunding campaigns. Firstly, the design of the crowdfunding platform allows the crowd to understand the purpose behind the project and the related services or products [3]. Secondly, the policies and norms of each crowdfunding platform affect the odds of the success of campaigns on the platform. This is explained by the study of Farnel [7], which suggests that Indiegogo's policies and norms were more receptive to gender reassignment campaigns compared to Kickstarter. Exploring the characteristics of crowdfunding platforms will provide an understanding of the impact it has on the success of crowdfunding projects. Additionally, the presence of other projects on the platform will influence the probability of success, such that campaigns with fewer competing projects will more likely receive funding [13].

However, crowdfunding participants are not always well-informed when choosing viable projects. When most investors are uninformed, they are inclined to mirror the decisions of well-informed investors who have a track record of supporting good projects [17]. Hence, information regarding previous contributors and investors can also potentially improve the conversion rate and amount of funding from subsequent funders [2]. To attract future users, crowdfunding platforms often include the fund raised by different campaigns on their website as well as the services including due diligence applications and additional services to signal investor protection.

As crowdfunding takes place online, unlike traditional financing, the likelihood of information asymmetry and imbalance increases. Therefore, it is crucial to examine the role of the crowdfunding platform in providing checks and due diligence prior to advocating campaigns. This process takes place before the project reaches the investors and acts as protection for funders. While most of the research focuses on the aspects such as the success of funders and ventures, it neglects the importance of the crowdfunding platform itself as a governance mechanism. In addition, inexperienced start-ups lead to higher risks borne by investors, and the relatively illiquid nature of equity crowdfunding further heightens the risk. Therefore, the effectiveness of crowdfunding intermediaries can serve to offset some of the risks and work towards preventing fraudulent behaviours [11]. Broker-dealers are subject to stringent regulations and due diligence standards imposed by the SEC and Financial Industry Regulatory Authority (FINRA), which ensure that investment decisions made by investors are rational by examining their risk profile and verifying the quality and authenticity of the issuers before listing their offering. However, apart from the basic due diligence check on the backgrounds of directors, officers, and equity holders of 20% or more, it is at the discretion of the funding portal whether they decide to take extra measures to verify the feasibility of the issuer and financial documents, or to provide risk management services to investors, to provide enhanced investor protection. Using the equity crowdfunding portals listed on FINRA, this study will examine the relationship between extra investor protection measures implemented by US equity crowdfunding platforms and the performance of funding portals as represented in the total amount raised on the portal.

2 Literature Review and Hypothesis Development

In this section, we conduct a literature review regarding various issues related to the equity crowdfunding portals and develop the hypothesis for our research.

2.1 Equity Crowdfunding Voting Rights

Portals adopt varying governance mechanisms with respect to voting rights offerings. For equity crowdfunding, voting rights are generally not given to smaller investors as venture capitalists and professional investors may be stirred away from investing in later rounds as it is troublesome and time consuming to collect signatures from all the crowd investors. Platforms can transfer voting rights to investors or to themselves directly, and they can also engage accredited investors or operate a nominee structure also sometimes known as pooled voting rights [19]. To attract venture capitalists and larger investors, some platforms also present the option for pooled investments where funding is collected from the crowd investors and the total amount is then invested as a single shareholder [20] to make follow-on financing easier to conduct. An example of this is the WeFunder portal which allows investors to pool their funds in the WeFund Special Purpose Vehicle and then invest it as one entity to the issuer. The WeFunder platform like a percentage of equity crowdfunding platforms is established as a public benefit company with the mission to contribute to the public in tandem with maximising shareholder value. As a result, these portals are likely to have more protection measures in place to uphold their reputation for the public good. In addition, some equity crowdfunding platforms also provide post-campaign support and services, and research implies that these follow-on services have a positive influence on the funding success rate [22].

2.2 Due Diligence and Investor Protection

The online nature of equity crowdfunding transactions and the involvement of non-professional investors can heighten investment risks, as it adds difficulty to the performance of cross checking and increases the chance of information asymmetric and moral hazard problems. Investor protection for small equity crowdfunding investors can be explored through three key avenues, i.e., legislation requirements, board governance, as well as investor protection measures taken by the crowdfunding platform.

Many countries have implemented rules regulating the maximum amount an investor can invest based on their incomes and wealth. For example, the JOBS Act in the US stipulates the investment threshold for investors under different Titles in relation to the amount they earn annually. In Germany, the Small Investor Protection Act requires investors to self-report their income and wealth for investments over €1,000, and only a corporate entity can invest more than €10,000 in a single equity crowdfunding issuer [9]. These legislations mainly provide protection from the aspect of ensuring their financial health is sufficient to support their investments and any associated risks. Once the investors have established that they have the capacity to fund equity crowdfunding issuers, the role of investor protection shifts emphasis to their own due diligence conducted on the firm, and the measures provided by the crowdfunding platform.

Unlike experienced venture capitalists and angel investors, individual investors often lack the means and incentive to conduct costly and time-consuming due diligence which can include background checks, credit checks, site visits and third-party verifications [5]. Hence, they rely on equity crowdfunding portals to list quality projects and exclude any risky campaigns and potential frauds. This has implications for the success of all parties. The study of Cumming et al. [5] on UK crowdfunding platforms found that the presence of due diligence is related to greater campaign success, more fund investors, and more capital raised, and these platforms often provide more additional services to both investors and project initiators. Similarly, the research of Rossi and Vismara [18] reveal that platforms that have a greater number of post-campaign services are correlated with a higher number of successful projects annually. As a result, these platforms are expected to have higher quality campaigns listed to enhance and maintain their reputation and appeal to more users.

2.3 Equity Crowdfunding Intermediaries

Equity financing intermediaries refer to funding portals or broker-dealers are the core facilitator matching investors with issuers in the equity crowdfunding process. These intermediaries provide a range of services and communication channels for issuers and investors to connect with one another. Under Title III of the 2012 JOBS Act, an issue must only provide equity offerings through an intermediary. As defined in the Act, funding portals are websites that are registered with the SEC and a member of FINRA that host a few equity crowdfunding activities and transactions.

This type of intermediary was newly created by the act, whereas broker-dealers are established market markers, and are permitted to give investment advice or suggestions to investors due to their extensive experience. Broker-dealer are faced with strict binding legislations in addition to due diligence responsibilities under the Know Your Customer Rule (FINRA 2090) and Suitability Rule (FINRA 2111) which stipulates that broker-dealers are required to collect information about each registered investor to determine their risk profile and which investment choices are appropriate [15]. Broker-dealers also have a set of clear due diligence standards that they are to comply with under the SEC and apply a number of criteria in the selection process of choosing which offerings to list to ensure the authenticity and quality of the issuer. Furthermore, they are required to implement reasonable measures to verify the accuracy and completeness of the information and document disclosed by the issuer. As such, investors are much more protected when engaging broker-dealers, however, this is accompanied by higher costs.

On the other hand, funding portals are not subject to such strict requirements, but they are still required to conduct background checks on directors, officers, and equity holders of 20% or more, to decrease the opportunity for fraud. In the case where one of the directors, officers, or a significant equity holder in the offering is deemed as a 'bad actor' by the definition of SEC (for example, a convicted felon, a person subject to

SEC disciplinary action, a person involved in a finance-related injunction or restraining order), the issuer must be disqualified.

Moreover, all funding intermediaries are subject to antifraud and antimanipulation provisions of the federal securities laws. Equity crowdfunding portals are not to have a financial interest in an issuer that uses their portal and must not compensate any third party leads to potential investors. Additionally, they are not permitted to receive, hold or manage any funds invested by the investor, and instead must adopt an independent escrow account that holds onto the funds and release it to the issuer only if the offering is successfully funded, other returns the funds to the investors.

As equity crowdfunding involves investment risks, it is pertinent that portals make investors aware of the potential risks prior to making their decision. Although these portals are not mandated to conduct extra due diligence besides the basic checks necessitated by SEC, most platforms urge investors to conduct their own thorough checks prior to making a funding decision, however, smaller investors lack the incentive to conduct due diligence due to the high cost. This may be revised based on campaign volumes, investor interest [6], and the platform's competitive position in the market.

Some platforms such as Crowdcube, go the extra mile and take steps either internally or with the partnership of a third party such as Creditsafe and Crowdcheck, to verify the offering and financial documents disclosed by the issuer and evaluate the viability of the business plan and purpose of operation before making a decision to accept or reject, in addition to providing educational resources and material to investors. Furthermore, these measures also have significant value for portals, as more funding success translates to a better reputation [20] and can likely attract more potential investors and issuers leading to more profit. Due diligence checks and processes are found to alleviate information asymmetry issues and improved the success of the portal in terms of the total amount raised [5]. A high level of due diligence can detect low-quality projects and entrepreneurs and reduce any reputation costs and litigation risks linked with the listing of such campaigns [5]. Furthermore, trust is an important factor that investors consider when selecting which campaigns and portals to place their investment in [12].

Since brokers and dealers are highly regulated and subject to more stringent due diligence standards and requirements, this study selects the equity crowdfunding portals registered on SEC and FINRA as the subject and examines the relationship between the level of investor protection and the total amount raised on the platform. In particular, we explore the different services targeted at investor protection, such as extra due diligence, selection criteria applied to issuers, educational resources, and a communication channel between the investor and issuer as facilitated by the portals, and whether the inclusion of these services in addition to the basic requirements by SEC will affect the performance of an equity crowdfunding portal as denoted by the total amount raised.

As funding portals are faced with less harsh expectations, it is rational to assume that some start-ups may deliberately approach funding portals instead of broker-dealers to take advantage of this, and hence it is anticipated that portals that have in place more extra services aimed at heightening investor protection will perform better. In addition, portals with a greater number of investor protection measures are likely to perform better. Hence, the following hypotheses are formed:

Hypothesis 1   Equity crowdfunding portals with extra investor protection measures in place will have a higher amount of total funds raised on the portal.

Hypothesis 2   Equity crowdfunding portals with a greater number of investor protection measures in place will have a higher amount of total funds raised on the portal.

3   Data and Methodology

This research takes insights from Cumming et al. [5] on due diligence in Canadian crowd-funding platforms, and Rossi and Vismara [18] on the role of investment-based platforms in Europe. However, this study uses data collected on US-based equity crowdfunding platforms as registered on FINRA.

3.1   Data Collection

To provide equity crowdfunding, a crowdfunding portal must be registered with the SEC as either a broker-dealer or a funding portal and must also be a member of FINRA. FINRA is a U.S. government-authorized not-for-profit organization that super-vises broker-dealers and funding portals to ensure the integrity of the securities market and protect investors. It operates under SEC oversight, developing rules and enforcing compliance with securities laws and FINRA regulations (FINRA 2021). Only registered funding portals and broker-dealers can offer and sell securities on behalf of issuers to public investors through crowdfunding. Issuers looking to make equity offerings can do so only through a registered intermediary.

The sample of U.S. equity crowdfunding portals was collected from the FINRA registration records as of December 2021, specifically from the Funding Portals We Regulate webpage. This page lists funding portal members of FINRA and those registered with the SEC. The initial sample included 78 equity crowdfunding portals, with each entry linking to the SEC's EDGAR database and the respective crowdfunding portal website.

Data on the total capital raised by each portal, up to December 2021, reflects the total funds raised since the portal's inception. Additional basic information, such as the number of employees, was gathered from the crowdfunding portal websites, as well as from sources like Crunchbase, Tracxn, and other company insights platforms.

The due diligence processes—including services offered to investors and any additional criteria for issuers—were collected manually from each portal's website. Supporting information about the crowdfunding platforms was obtained from documents submitted to the SEC. Due to the limited availability of data on equity crowdfunding platforms, the final sample comprised 53 equity crowdfunding portals after excluding entries with incomplete data for all variables used in the regression model.

The table below details the description of the dependent, independent, and control variables (Table 1).

This table shows the list of variables including the notation and description of the measurement of each variable used in the regression. The dependent variable is the total amount of capital raised on the crowdfunding portal (in millions). The independent variables include the extra due diligence, education resources for investors, a communication

Table 1. List of variables.

Variable	Variable notation	Description
Dependent variable		
Total amount raised	*Amount*	The total amount of capital raised on the crowdfunding platform (in millions)
Independent variables		
Presence of extra due diligence	*E_DueD*	Dummy variable: equals 1 if the crowdfunding platform conducts extra due diligence, and 0 if otherwise
Education for investor	*Edu*	Dummy variable: equals 1 if the crowdfunding platform provides educational material or resources to investors, and 0 if otherwise
Communication channels with issuers	*Comms*	Dummy variable: equals 1 if the crowdfunding platform facilitates communication between investors and issuers, and 0 if otherwise
Criteria	*Criteria*	Dummy variable: equals 1 if the crowdfunding platform imposes selection criteria of issuers, and 0 if otherwise
The number of investor protection services	*N_InvP*	The number of investor protection measures present on the crowdfunding platform as denoted by the presence of extra due diligence, educational resources for investors, a communication channel between the investors and issuers and selection criteria of issuers. 0 is the lowest value and 4 is the highest value
Control variables		
Platform age	*PAge*	Years since the platform was founded
Other fees charged by the portal	*Fees_Other*	Dummy variable: equals 1 if the crowdfunding platform charges other fees including listing fee, and other service fees, and 0 if otherwise
The number of employees	*N_Employees*	The number of employees working for the crowdfunding platform. This is a dummy variable: equals 1 if over 100 employees and 0 otherwise (i.e., less than or equal to 100 employees)

between investors and issuers, selection criteria and the number of investor protection measures. The control variables include platform age, other fees charged by the portal as well as the number of employees working for the portal.

3.2 Dependent Variable

The dependent variable (Amount) of this study is the total amount raised to date (2021) as collected from Crunchbase. It is a proxy for the performance of the equity crowdfunding portal, as a higher value suggests better performance since more investors are choosing to invest in the offerings listed on the portal. Drawing from the findings of the study of Cumming et al. [5], platforms that provide due diligence services are associated with higher rates of successful campaigns and the total amount raised.

Due to the data limitations and lack of data available, this paper focuses only on the total amount raised. However, when data on the number of successful campaigns on each US equity crowdfunding portal become available, this research can be extended to test the impact of extra due diligence and investor protection measures on the number of successful campaigns.

3.3 Independent Variables

Expanding on the study of Cumming et al. [5], this paper does not focus on whether due diligence is conducted as all funding intermediaries are required by SEC and FINRA to perform basic background checks of directors, officers and equity holders of 20% or more, as well as disqualify any 'bad actors' as deemed by the definition used by SEC. Instead, we examine whether extra due diligence processes and investor protection measures are present in the forms of due diligence checks in addition to the basic checks as required (E_DueD), educational resources and materials (Edu) to educate investors about equity crowdfunding and the risks involved, a communication channel (Comms) between the investors and the issuer as facilitated by the funding portal, the extra selection criteria imposed on the issuers by the portal (Criteria) and the number of investor protection measures (N_InvP) in place. It is expected that portals with extra and more measures present will lead to a great total amount raised, since it heightens the chance of listing quality campaigns and signals better platform management and governance and investor protection to potential funders.

3.4 Control Variables

Referring to the study of Cumming et al. [5] in tandem with other studies on crowdfunding platforms and the availability of data, this study uses the platform age, other fees charged by the equity crowdfunding portal (including fixed percentage, listing fee and other service fees), the number of employees working for the funding platform, and the industry the platform mainly focuses on as the control variables. The data contains only the range of the number of employees for each platform rather than the exact number, therefore the variable for the number of employees is a dummy variable, whereby it takes 1 if the equity crowdfunding portals has more than 100 employees and 0 if it has less than or equal to 100 employees. The industry variable is not included, as the equity crowdfunding portals in the sample generally accept all types of campaigns rather than strictly targets a certain sector.

3.5 Regression Model

In this paper, we use the multiple linear regression analysis to examine the association between investor protection measures (e.g. extra due diligence checks, educational resources provided to investors, additional criterions set by the funding portal and a communication channel between investors and the issuer as hosted by the portal) and the performance of the equity funding portal as represented by the total amount raised (in millions of dollars).

This model tests the relationship between the extra due diligence processes (*E_DueD*), educational resources (*Edu*), communication channel (*Comms*) selection criteria (*Criteria*), and the number of investor protection measures (*N_InvP*) on equity crowdfunding portal performance as denoted by the total amount raised (*Amount*). The control variables consist of *PAge*, *Fees_Other*, and *N_Employees*, representing the age of the portal, whether the portal charges other fees and the number of employees of the portal, respectively. Hence, the regression models can be expressed as below:

$$Amount_i = \beta_0 + \beta_1 E\_DueD_i + \beta_2 Edu_i + \beta_3 Comms_i$$
$$+ \beta_4 Criteria_i + \beta_5 N\_InvP_i + \beta_6 PAge_i + \beta_7 Fees\_Other_i$$
$$+ \beta_8 N\_Employees_i + \varepsilon_i$$

where $Amount_i$ represents the firm of the equity crowdfunding portals as proxied by the total amount raised on the platform, β_0 is the constant, β_{1-5} and β_{6-8} are the slopes of the independent and control variables, respectively, and ε is the error term.

4 Sample and Empirical Findings

4.1 Descriptive Statistics

Table 2 below presents the descriptive statistics of the sample which consists of 53 observations. The mean for the variable *Amount* is 10.2143, indicating the mean value for the total amount raised is USD10.2143 million and ranges between 0.002 million to 214 million. The standard deviation of 4.5672 indicates a large spread. The mean for *E_DueD*, *Edu*, *Comms*, and *Criteria* is 0.4528, 0.879, 0.7547, and 0.2075 respectively, implying, just under half of the equity crowdfunding portals have extra due diligence processes, many portals provide educational resources and facilitate communication channel between investors and issuers, while a small percentage of approximately 20% of the portals impose their own selection criteria when screening projects. The mean for the variable *N_InvP* is 2.1321, and ranges between 0 and 4, indicating that in general, equity crowdfunding portals apply 2 investor protection measures. The average age of the portals is 4.4528 years, with less than 100 employees on average. The variable for other fees has a mean of 0.9434, showing that most equity crowdfunding platforms charge other fees such as a fixed percentage fee for the amount raised, a listing fee and other fees for other services.

Table 2. Descriptive statistics.

Variable	Obs.	Mean	Std. Dev.	Min	Max
Amount	53	10.2143	4.5672	0.002	214
E_DueD	53	0.4528	0.0690	0	1
Edu	53	0.8679	0.0470	0	1
Comms	53	0.7547	0.0597	0	1
Criteria	53	0.2075	0.0562	0	1
N_InvP	53	2.1321	0.1476	0	4
PAge	53	4.4528	0.4019	1	13
Fee_Other	53	0.9434	0.0320	0	1
N_Employees	53	0.0377	0.0264	0	1

This table presents the descriptive statistics (mean, standard deviation, min, and max) for the sample of 53 U.S. equity crowdfunding portals. *Amount* (total amount raised) is the dependent variable, and *E_DueD* (extra due diligence), *Edu* (educational resources for investors), *Comms* (communication channel between investors and issuers facilitated by the portal), *Criteria* (selection criteria) and *N_InvP* (number of investor protection measures) are the independent variables. *PAge* (platform age), *Fees_Other* (other fees), and *N_Employees* (number of employees working at the portal) are the control variables.

4.2 Multicollinearity

When multicollinearity is present, the regression results may be spurious. To check the existence of multicollinearity, a correlation matrix is presented in Table 3. It shows that most of the variables are weakly correlated except three pairs of variables: *Amount* and *N_Employees*, *E_DueD* and *N_InvP*, and *Comms* and *N_InvP* which have correlation coefficients of 0.8435, 0.6348 and 0.6474, respectively. The positive and high linear correlation between the total amount of funds raised and the number of employees working at the portal is expected, as it is reasonable to assume that the number of employees required by the portal is associated to the funds raised as likely to be reflected in the number of campaigns listed. Likewise, on the other hand, equity crowdfunding portals with a higher workforce capacity can potentially process more campaigns and funding projects. From the correlation matrix in Table 3, it is evident that the extra due diligence, education resources, communication channels and selection criteria have a positive and highly significant (at 1% level) correlation with the number of investor protection measures implemented by the equity crowdfunding portal. This may be explained by the inference that portals that are willing to implement extra measures will be more likely to adopt more than one to ensure the quality of the campaigns and issuers on their portal which will be reflected in the reputation and credibility of the portal. In addition, the selection criteria and extra due diligence also have a positive and highly significant (at 1% level) correlation.

It is worth noting that estimating variance inflation factor (*VIF*) provides a more accurate assessment of collinearity issues compared to calculating correlation coefficients. A rule of thumb is that if $VIF > 10$, the multicollinearity is high [14]. The estimation reveals that none of the VIF values exceeds 10, which indicates the absence of significant multicollinearity concerns in the analysis.

In sum, Table 3 shows that the multicollinearity issue is not significant with the linear regression model.

Table 3. Correlation coefficient matrix and VIF.

	Amount	E_DueD	Edu	Comms	Criteria	N_InvP	PAge	Fee_Other	N_Employees	VIF
Amount	1									
E_DueD	0.127	1								2.301
Edu	0.102	0.131	1							2.800
Comms	0.143	0.254*	0.166	1						3.131
Criteria	-0.009	0.376***	-0.075	0.075	1					2.256
N_InvP	0.091	0.635***	0.519***	0.647***	0.461***	1				9.159
PAge	0.129	-0.194	0.099	-0.214	-0.016	-0.178	1			1.176
Fee_Other	0.074	-0.105	0.146	0.24*	-0.076	0.107	0.151	1		1.151
N_Employees	0.844***	0.019	0.077	0.113	-0.101	0.068	0.037	0.049	1	1.035

This table contains the correlation coefficient matrix which is performed to examine to association between the variables used in the regression model to detect the presence of multicollinearity. In addition, the variance of inflation factor (VIF) is also reported for each independent variable. *, ** and *** indicate significance at the 10%, 5%, and 1% levels, respectively.

4.3 Regression Results

The regression findings are reported in Table 4 below. The R squared value of 0.7764 indicate a strong fit between the dependent variable and the explanatory variables. It suggests that almost 78% of the movement in *Amount*, is explained by the model. The adjusted R square presents a better representation of the model fit as it accounts for the addition of the new variables. It is evident that when only the control variables were included in the model as shown in column (1), the adjusted R square had a value of 0.7042, and when the independent variables were added to the model as shown in column (6), the adjusted R square value increased to 0.7357 indicating that the inclusion of the independent variables increased the model fit. The adjusted R squared of 0.7357 demonstrate that 73.57% of the variation in the dependent variable (Amount) is defined by variations in the 5 independent variables, implying a strong model fit and that the addition of independent variables are relevant to the increase in model fit. Furthermore, the F-stats value of 19.0974 is greater than the critical value of 2.9457 as calculated using the F-distribution table and the p-value of the test as denoted by the Sig.F value is 0, signifying that there is great evidence to infer that the model is valid.

The coefficient for the variable *E_DueD* is 19.7623, indicating that as extra due diligence increases by one unit, the dependent variable (*Amount*) is expected to increase by 19.7623 units on average. The variable for *Edu* has a coefficient value of 23.9362, suggesting that as educational resources for investors increase by one unit, the dependent variable (*Amount*) is expected to increase by 23.9362 units on average. The coefficients of *Comms* and *Criteria* show that if the variables increase by one unit, *Amount* is expected to increase on average by 22.5108 and 18.1453 units on average, respectively. The variable *N_InvP* has a negative coefficient value, demonstrating that when the number of investor protection measures increase by one unit, the total amount raised on the portal decreases by 17.4272 units on average. This implies that on the one hand, imposing investor protection measures increases the total amount of funds raised as it signals emphasis on the quality of crowdfunding campaigns and attracts investors. But, on the other hand, the increase in the number of investor protection measures will decrease the total amount raised as conditions that are too stringent are likely to repel issuers hence reducing the attractiveness of the portal for issuers and initiators of crowdfunding campaigns, and consequently the number of campaigns listed on the portal.

The variables of *E_DueD, Edu, Comms, Criteria* are all positive and statistically significant at the 1%, 5%, 5% and 5% levels, respectively, while the variable *N_InvP* is negative and statistically significant at the 5%. This demonstrates that extra due diligence checks, educational resources, a communication channel facilitated by equity crowdfunding portals and selection criteria all have a positive association with the total amount raised on the platform. This supports Hypothesis 1 stating that equity crowdfunding platforms with extra investor protection measures perform better in terms of the total amount raised.

However, the number of investor protection measures is negative and significant at the 5% level, this may suggest that platforms with more measures will be less attractive to issuers and hence deter the likeliness of them choosing the platform. This finding, therefore, rejects Hypothesis 2 stating that a greater number of investor protection measures employed by the equity crowdfunding platform will lead to better performance. The variable *N_Employees* is positive and statistically significant at the 1% levels, implying that as the number of employees increases, the total amount raised will also increase. This can be explained by the fact that more employees support the implementation of extra investor protection measures, which suggests better quality equity crowdfunding campaigns are listed on the portal, in turn drawing more investors to fund the projects, and leading to more funds raised on the platform.

This table shows the regression findings on the research model with a sample of 53 observations. Column (1) shows the findings when only the control variables are included in the model. The independent variables are added one by one in Columns (2) to (6) to capture the impact of the addition of the independent variables. *, **, *** represent the significance level of 10%, 5% and 1%, respectively. The standard errors are reported in the parenthesis.

Table 4. Regression results.

Model	(1)	(2)	(3)	(4)	(5)	(6)
Intercept	-2.5819	-9.4461	-9.6307	-10.2433	-10.7143	-16.1899
	(10.6932)	(11.0951)	(11.9602)	(12.1279)	(12.2975)	(11.7605)
E_DueD		9.1427*	9.1054*	8.5146	7.6358	19.7623***
		(4.9862)	(5.1087)	(5.2980)	(5.7849)	(7.1553)
Edu			0.3307	-0.1052	0.3132	23.9362**
			(7.4617)	(7.5791)	(7.7212)	(11.6030)
Comms				3.0658	2.9792	22.5106**
				(6.4270)	(6.4903)	(9.6563)
Criteria					2.6951	18.1453**
					(6.7702)	(8.6945)
N_InvP						-17.4272**
						(6.6737)
PAge	1.0732	1.3616	1.3575	1.4606	1.4263	1.0616
	(0.8674)	(0.8617)	(0.8758)	(0.9092)	(0.9217)	(0.8785)
Fees_other	2.6977	4.2521	4.1825	2.6032	2.7537	3.0320
	(10.8835)	(10.6641)	(10.8907)	(11.4695)	(11.5821)	(10.8994)
N_Employees	145.0181***	144.3145***	144.2774***	143.6190***	144.2187***	145.5900***
	(10.8835)	(12.7581)	(12.9200)	(13.1004)	(13.3075)	(12.5335)
R square	0.7213	0.7395	0.7396	0.7408	0.7417	0.7764
Adjusted R square	0.7042	0.7178	0.7118	0.7070	0.7016	0.7357
F stats	42.2723	34.0730	26.6920	21.9157	18.4639	19.0974
Sig.F	0.0000	0.0000	0.0000	0.0000	0.0000	0.0000
Obs	53	53	53	53	53	53

5   Concluding Remarks

The online nature of equity crowdfunding, combined with the participation of non-professional investors and the inherent high risk of equity returns, can intensify issues such as hidden information, adverse selection, and information asymmetry. To enhance investor protection and improve regulation in this sector, the JOBS Act imposes limits on the maximum amount a non-accredited investor can contribute, based on their annual income or net worth. This ensures that investors have adequate financial capacity relative to their investment choices. Furthermore, both the SEC and FINRA mandate that all equity crowdfunding intermediaries—including broker-dealers and crowdfunding portals—conduct background checks on directors, officers, and significant equity holders (those owning 20% or more) and disqualify any individuals labelled as "bad actors" by the SEC.

While broker-dealers are subject to stringent due diligence standards, equity crowd-funding portals have discretion over whether to implement additional checks or services to bolster investor protection. Although these portals are required to perform basic due diligence checks on issuers before listing their campaigns, this paper investigates the relationship between investor protection measures—such as enhanced due diligence checks, educational resources, communication channels, and additional selection criteria—and platform performance, as indicated by the total amount raised.

The findings reveal that enhanced due diligence processes, educational resources, effective communication channels, and stringent selection criteria significantly positively influence the total amount raised. Conversely, an increased number of investor protection measures appears to have a negative effect. Additionally, a higher number of employees correlates positively with the total amount raised, which is expected since portals that implement robust investor protection measures likely require more personnel to manage the increased workload. This suggests that more employees facilitate the implementation of these measures, leading to higher-quality equity crowdfunding campaigns being listed on the portal, thereby attracting more investors and ultimately increasing funds raised.

These findings carry important implications for equity crowdfunding portals, regulators, investors, and issuers, highlighting the significance of enhanced due diligence and investor protection measures on platform performance, issuer success rates, and investor returns. Equity crowdfunding platforms should proactively consider implementing additional due diligence checks, as this could positively impact their performance, reputation, and credibility, thereby attracting more users in the future.

However, portals must exercise caution regarding the quantity of protective measures implemented, as an excessive number may deter issuers and diminish the portal's competitiveness in attracting crowdfunding campaigns. Regulators might also consider instituting stricter regulations for equity crowdfunding portals to bolster both performance and investor protection. Investors and issuers can use these insights to identify measures that contribute to greater success in the crowdfunding space. Additionally, these measures can serve as signals to investors regarding the quality of the campaigns available on the portal, while still remaining appealing to issuers.

Finally, the results of this paper should be interpreted with caution due to the limited sample size employed, which could be addressed in future research.

References

1. Belleflamme, P., Lambert, T., Schwienbacher, A.: Crowdfunding: tapping the right crowd. J. Bus. Ventur. **29**(5), 585–609 (2014)
2. Burtch, G., Ghose, A., Wattal, S.: An empirical examination of the antecedents and consequences of contribution patterns in crowd-funded markets. Inf. Syst. Res. **24**(3), 499–519 (2013)
3. Choy, K., Schlagwein, D.: Crowdsourcing for a better world: On the relation between IT affordances and donor motivations in charitable crowdfunding. Inf. Technol. People **29**(1), 221–247 (2016)
4. Cumming, D.J., Johan, S.A.: Crowdfunding: fundamental cases, facts, and insights. 1st edn. Academic Press (2019)
5. Cumming, D.J., Johan, S.A., Zhang, Y.: The role of due diligence in crowdfunding platforms. J. Banking Financ. **108**, 105661 (2019)

6. Cumming, D.J., Vanacker, T., Zahra, S.A.: Equity crowdfunding and governance: toward an integrative model and research agenda. Acad. Manag. Perspect. **35**(1), 69–95 (2021)

7. Farnel, M.: Kickstarting trans*: the crowdfunding of gender/sexual reassignment surgeries. New Media Soc. **17**(2), 215–230 (2015)

8. FINRA Funding Portals We Regulate FINRA. https://www.finra.org/about/firms-we-regulate/funding-portals-we-regulate. Accessed 18 Sep 2021

9. Goethner, M., Hornuf, L., Regner, T.: Protecting investors in equity crowdfunding: an empirical analysis of the small investor protection act. Technol. Forecast. Soc. Chang. **162**, 120352 (2021)

10. Johan, S., Zhang, Y.: Quality revealing or overstating? analysis of qualitative startup information in equity crowdfunding. Working Paper, Florida Atlantic University and Gonzaga University (2018). https://ssrn.com/abstract= 3291905

11. Kamalnath, A., Lin, N.: Crowd-sourced equity funding in australia — a critical appraisal. Fed. Law Rev. **47**(2), 288–305 (2019)

12. Liang, T.P., Wu, S.P.J., Huang, C.C.: Why funders invest in crowdfunding projects: role of trust from the dual-process perspective. Inf. Manag. **56**(1), 70–84 (2019)

13. Meer, J.: Effects of the price of charitable giving: evidence from an online crowdfunding platform. J. Econ. Behav. Organ. **103**, 113–124 (2014)

14. Neter, J., Wasserman, W., Kutner, M.H.: Applied linear regression models, 2nd edn. Irwin, Richard D (1983)

15. Nutting, M.R., Freedman, D.M.: Equity crowdfunding for investors: a guide to risks, returns, regulations, funding portals, due diligence, and deal terms. John Wiley & Sons (2015)

16. Ordanini, A., Miceli, L., Pizzetti, M., Parasuraman, A.: Crowd-funding: transforming customers into investors through innovative service platforms. J. Serv. Manag. **22**(4), 443–470 (2011)

17. Parker, S.C.: Crowdfunding, cascades and informed investors. Econ. Lett. **125**(3), 432–435 (2014)

18. Rossi, A., Vismara, S.: What do crowdfunding platforms do? a comparison between investment-based platforms in Europe. Eurasian Bus. Rev. **8**, 93–118 (2018)

19. Rossi, A., Vismara, S., Meoli, M.: Voting rights delivery in investment-based crowdfunding: a cross-platform analysis. J. Ind. Bus. Econ. **46**, 251–281 (2019)

20. Schwienbacher, A.: Equity crowdfunding: anything to celebrate? Ventur. Cap. **21**(1), 65–74 (2019)

21. Schwienbacher, A., Belleflamme, P., Lambert, T.: Individual Crowdfunding Practices. Ventur. Cap.: Int. J. Entrepreneurial Financ. **15**(4), 313–333 (2013)

22. Yasar, B.: The new investment landscape: equity crowdfunding. Central Bank Rev. **21**(1), 1–16 (2021)

Financial Time Series Simulation with Transformer-Based Generative Models Under Continuous Conditions

Horstann Rui Yao Ho[1] and Chi Seng Pun[2(✉)]

[1] College of Computing and Data Science, Nanyang Technological University,
Singapore 639798, Singapore
[2] School of Physical and Mathematical Sciences, Nanyang Technological University,
Singapore 637371, Singapore
cspun@ntu.edu.sg

Abstract. Computer-generated time series simulations have been heavily used in banks and hedge funds for risk management, pricing, volatility trading, hedging, etc. Traditionally, these simulations require binding model-based assumptions about how market prices move, which often may not be reflected in real-life, and they usually fail to capture either fat tails or non-symmetric distributions. To address this shortcoming, we propose a novel generative deep learning frameworks, built on top of generative adversarial networks (GANs) given their strong capability in simulating synthetic data, called continuous conditional transformer-based time-series GANs (CC-TTS-GAN). CC-TTS-GAN possess the ability to incorporate some market conditions (such as implied volatility or other market indicators) which are assumed to be continuous in nature, to inform their own simulations in a model-free manner. We then compare our model's simulations to traditional benchmark models (calibrated to real data) such as geometric Brownian motion, constant elasticity of variance model and Heston stochastic volatility model, based on a wide range of evaluation metrics. The outperformance of our proposed CC-TTS-GAN over the model-based benchmarks is statistically significant.

All implementation code can be found at https://github.com/Horstann/CC-TTS-GANs.

Keywords: Financial Time Series · Simulation Techniques · Deep Learning · Generative Adversarial Networks (GAN) · Transformer · Conditional GAN

1 Introduction

Nowadays, machine learning techniques have proliferated throughout the banking and finance industry in applications such as investment insights, alpha generation, risk management, derivative pricing and more. This is an exciting and promising area with ongoing research by institutions and asset managers who strive to capture these nascent opportunities to stay ahead in the financial markets.

© The Author(s), under exclusive license to Springer Nature Singapore Pte Ltd. 2025
K.-W. Huang et al. (Eds.): ICFT 2024, CCIS 2437, pp. 191–206, 2025.
https://doi.org/10.1007/978-981-96-3811-6_17

This paper aims to explore innovations within the sub-area of Monte-Carlo simulations for time-series data, using a novel variation of generative adversarial networks (GANs). Monte-Carlo simulations are extremely useful in industry, as they can be used to interpret the distributions and other statistical properties of returns or price paths. These are critical in modelling sensitivity, pricing complex exotic options, forecasting, as well as analyzing quantiles for methods such as value-at-risk and expected shortfall.

GAN models, proposed in [4], have been noted for their amazing ability of generating realistic and high-quality images. The usual GAN architecture involves a generator model that aims to generate images indistinguishable from real images and a discriminator (or critic) model that aims to distinguish between real and generated images. Both models are usually constructed using deep neural networks. Such an adversarial dynamic allows us to train both the generator and discriminator models to be experts in their respective objectives in an unsupervised setting. In the past, GAN applications have been focused on image generation and text generation. Only recently, we observe more publications on how this framework could instead be used to generate time-series simulations.

However, we note that many existing GANs for time-series simulations are unconditional, which prevents the incorporation of conditions (prior to simulations) that are often crucial in capturing context the financial markets. These conditions could be important input variables such as market indicators (e.g. RSI, market sentiments) and option chains data (e.g. volatility, option prices, put-call ratios), which may heavily influence the distribution of time series simulations. Besides, traditional GANs for images trained on conditional generation often only use categorical conditions (e.g., "cat" and "dog"). This poses a concern if our conditions are continuous in nature, such as our aforementioned financial variables.

Hence, through this paper we aim to contribute valuable insights within this area of research by developing a GAN model that is capable of modelling the distribution of time-series simulations given some continuous conditions – a continuous, conditional GAN for financial time series. This capability will lead to more accurate forward-looking distributions of time-series data, from which users can extract more meaningful and valuable insights in an industry setting.

1.1 Related Works

There have been many experiments in leveraging the powerful capabilities of the traditional GAN framework. Each offering different benefits and insights with respect to this paper's goal.

[8] showcased how GANs can be very well adapted to time-series data and produce realistic simulations when coupled with the transformer mechanism. However, generations from this model were unconditional so users could not incorporate any information prior to simulation. Moreover, experiments were mainly performed on non-financial time series such as random sinusoidal waves, human heartbeat signals, and other human activity recognition data. Much of these datasets were stationary and non-trending, unlike financial price paths.

On the other hand, [3] was the first to introduce a GAN with a new vicinal loss function that enabled the incorporation of continuous conditions to generate images, such as objects at specified angles and human faces at specified ages. It was made possible by sampling training examples within the vicinity of a target condition. These results were promising and opened further possibilities for incorporating conditions in time series generation. However, in this setup, we must assume that our conditions are relatively uniformly distributed, and we have sufficient number of samples. This can be true for object angles and human ages, but not so for financial market variables.

Another interesting work by [11] proposed implementing GANs focused on generating financial time-series, but under a semi-supervised setting. The generator model was trained on a supervised economics-based loss function with terms that measure profit and loss, mean square error, and Sharpe ratio.

Our paper attempts to draw upon ideas from these various pieces to construct a GAN model capable of achieving our objectives. We aim to show that such a model has the potential to add value beyond what existing models are capable of and can be extended across different areas of research or applications within quantitative finance.

2 Data Collection

In this paper, we will experiment with simulations of the S&P500 ETF end-of-day price paths, as data available for the ETF and its options are most freely available. It will also be a good starting point before we extend our model to other ETFs, single-stocks, or even other asset classes. We were able to retrieve S&P500 index prices from Jan 2005 to present day easily using the Yahoo!Finance API.

As for the determination of market conditions, we consider variables such as options implied volatility, options put-call ratio, and relative strength index (RSI) of S&P500 index prices. Implied volatility data were extracted from the Bloomberg LIVE (Listed Implied Volatility Engine), that provided the daily implied volatility of listed options for all global asset classes back until Jan 2005, given days to expiry and moneyness. We extracted the implied volatility of S&P500 at-the-money options with days to expiry of 30 and 60. The end-of-day RSI was easily computed directly from the S&P500 price paths.

Lastly, we also require end-of-day market prices of S&P500 options to calibrate some of the Monte-Carlo benchmark models, namely the constant elasticity of variance (CEV) and Heston models. An open source of historical end-of-day options chain data going back until Jan 2005 can be found at OptionsDx.com. Once retrieved, for each day, we filtered for options with expiry of within 30 and 90 days, and then computed the mean moneyness level (strike price divided by underlying price) and average market price for all put and call options. With mean strike price, underlying price, mean days to expiry and option market price, we are now able to calibrate parameters for our Monte-Carlo benchmarks.

3 Methodology

The objective of our experiment is to produce forward-looking distributions of daily S&P500 closing price paths within 2-month windows.

Our time series from Jan 2005 to present day will have a train-test split of 90–10. The training period is from Jan 2005 to Apr 2022 and the testing period is from Apr 2022 to

May 2024. The training size of 90% is to allow for a larger sample size to draw training examples from, and to include as much of the volatile Covid-19 pandemic period in training. The model will be trained on the whole training set for multiple epochs, and then evaluated on the whole test set at once. As we do not use a rolling model that is retrained at each timestep, this renders our problem more challenging and leaves some room for potential improvements. Despite so, we will show that the conditioning capability of our framework allows us to generate sensible price paths.

3.1   Preprocessing and Feature Engineering

Before constructing the model, we first perform some necessary data pre-processing to ensure our data can be fitted by the model.

First, we calculate the log-returns of the S&P500 price paths. Log-returns are more stationary in nature and thus easier to model, and it could also allow the model to better capture volatility clusters. Most importantly, as we will cover later, the GAN model will perform layer normalization at each network layer and thus its output will almost always be stationary. After applying a Dickey–Fuller test on the log-returns, we obtain a p-value of 0.00, which indicates that the log-returns can be deemed stationary.

Besides that, we also normalize our conditions using a rolling z-score, which allows all our conditions to be on the same scale that is essential in sampling vicinal examples with respect to a target condition using L2 distance, as we will see later. The rolling window we set for each variable is around 1 year or less, determined by the p-values of Granger causality tests applied across 2-month windows, where the treatment is the standardized values and outcome are the log-returns.

Table 1 summarizes Granger causality test results of our standardized variables when applied on log-returns and close prices. We see that the p-values for log-returns are substantially lower than that for close prices, which proves our point that log-returns are easier to model. As we observe relatively low mean p-values for these three variables, this indicates that they have some predictive potential over log-returns. Hence, we will proceed to use them in our condition embeddings. Though, note that we excluded put-call ratio from our condition variables as it was too noisy and would unlikely be a meaningful signal for an output window of two months.

Table 1. p-Value Statistics from Granger Causality Tests

Variables		RSI	30d ATM implied vol	60d ATM implied vol
Z-score rolling window size		252	189	189
Causality on log-returns	Min p-value	0.0029	0.0015	0.0000
	Mean p-value	0.1035	0.0563	0.0044
Causality on close prices	Min p-value	0.0736	0.0197	0.0004
	Mean p-value	0.3517	0.3921	0.0489

3.2 Model Architecture

Our CC-TTS-GAN's model architecture is composed of the usual generator and discriminator. The generator takes in a noise vector and condition embedding to generate a 2-month log-returns series. This is converted back to a price path that has an initial value of 1.0, which in turn is fed alongside the same condition embedding into the discriminator, which produces a score that reflects the "realness" of the generator's output. The overview of the model architecture can be referred in Fig. 1, which builds on top of the architecture figure in [8].

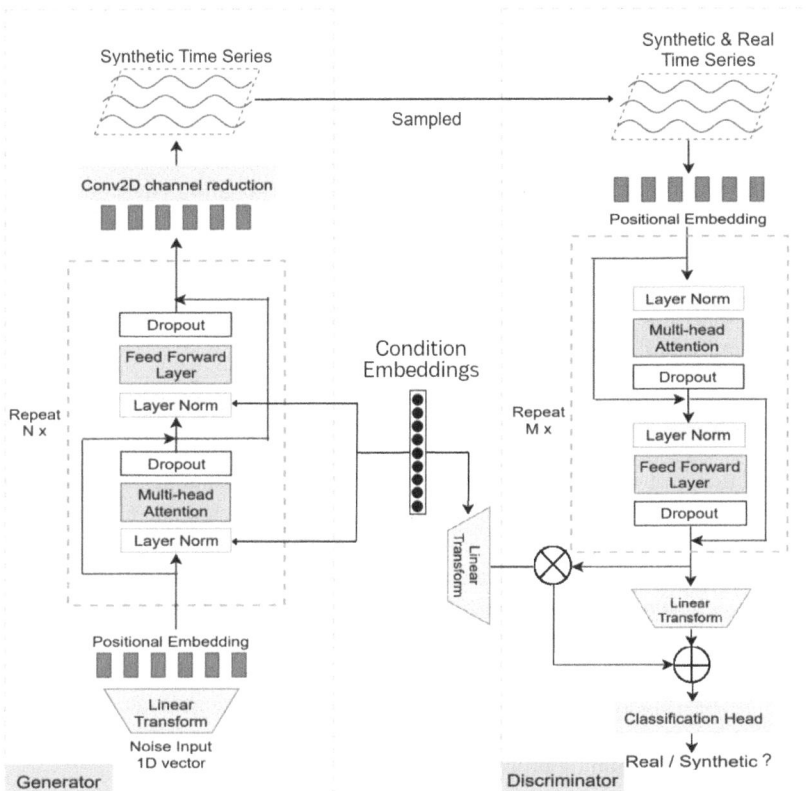

Fig. 1. Architecture of CC-TTS-GAN

There are two aspects to the architecture that are essential to creating a CC-TTS-GAN, which are detailed below.

3.2.1 Aspect 1: Transformer-Based Time-Series Architecture

We first establish an unconditional GANs model that leverages on the transformer architecture. Transformers and their attention mechanisms have grown in popularity as they have proven to outperform previous neural network architectures commonly used in GANs, such as RNNs, LSTMs and CNNs; see [10]. This is attributed to their ability to handle long sequences without suffering from gradient vanishing, which is extremely useful in time series analysis.

The time series is divided and processed in patches similar to how the vision transformer processes images; see [7]. Both the generator and discriminator model start with positional encoding, followed by three self-attention layers. The generator also has five attention heads in each transformer layer to allow for more complex computations and more diverse outputs. Our architecture in this aspect is similar to that of [8] while we follow the same positional embeddings as [8] suggested. Note that our positional embeddings are randomly generated vectors produced at the start of training.

3.2.2 Aspect 2: Conditional Generation

Extending from the architecture in Aspect 1, we intend to incorporate prior conditions. Simply concatenating the condition embeddings our inputs is the simplest and crudest approach. However, feeding them at every layer of the generator allows the conditions to more strongly influence the generated output and produce more diverse results. Hence, we adopt a two-dimensional conditional layer normalization after each self-attention layer.

As for the discriminator, we simply perform an add-norm operation to incorporate our condition embeddings into the final discriminator layer.

3.3 Training Methodology

In this subsection, we explain how we can sample training examples from a target condition, which is an important step in allowing us to train the model to achieve robust and accurate conditional generation. We then go over how we use these real and fake examples to update the weights of the discriminator and the generator.

We introduce a supervised generator loss function component on top of the unsupervised GAN dynamic that measures the quality of an entire distribution of generated price paths given a single condition embedding, which is very beneficial since the discriminator only measures the quality of each individual price path generated rather than the entire distribution as a whole. Note that Kaggle GPU resources (GPU T4 x2) were utilized to speed up the training process.

3.3.1 Vicinal Sampling Given Target Condition

In each training epoch, we iterate through each training example's target conditions. Each target condition will be added with a small amount of Gaussian noise. The rule of thumb we use for the standard deviation of this Gaussian noise is

$$1.06 \times \text{Stdev\_condition} \times (\text{size\_train})^{\wedge}(-1/5)$$

As proposed by [3]. This is important in ensuring that our input conditions during training are not just fixed to those available in the training set and makes outputs smoother and more robust especially for input conditions that have not been seen by the generator before.

Once done, we use L2 distance to obtain the top 10 training examples whose conditions are most similar to these new conditions embedding with noise added in. This will be the real price paths shown to the discriminator. The generator on the other hand will also be fed these new conditions embedding and tasked to generate 30 examples, 10 of which will be randomly selected and fed into the discriminator. In this way, we can ensure the discriminator sees an equal number of real and fake examples, while having a localized selection of 10 real examples within the very immediate vicinity of the target condition. At the same time, we have a large enough sample of 30 fake examples, which we can use our supervised loss function for further evaluation.

We use 40 epochs and a batch size of 4, as this was found to be the most optimal. To clarify, each item in the batch refers to one unique condition. As such, a batch size of 4 means 4 unique conditions, which in turn means $4*10 = 40$ real examples and $4*30 = 120$ fake examples per batch.

3.3.2   Update Rules for Generator and Discriminator

Once we have collected a few real and fake examples, these can be used to train and update the weights of the discriminator and generator models in a two-step process.

Step 1: Update the discriminator. This involves updating only the discriminator, which in this case, works like a critic. Instead of acting like a classifier that assigns binary or probit scores between 0 and 1 for fake and real, it scores each example with a value. This value can be positive or negative while what matters is that higher values reflect a higher degree of "realness". This is essentially the Wasserstein loss function (see [1]) as follows:

$$disc\_loss = mean(scores\_false) - mean(scores\_real) + gradient\_penalty$$

Note that we also apply a gradient penalty that forces the norm of gradients to be close to 1. This ensures the 1-Lipschitz continuity of the discriminator gradients, which helps the discriminator train better when using the Wasserstein loss; see [1]. For each condition data point, the loss function above will be applied on 10 real examples and 10 fake examples (randomly selected from the 30 fake examples).

Step 2: Update the generator. This involves updating only the generator. Through empirical experiments, we find that using the usual Wasserstein loss on the generator of just mean (scores_false) could sometimes lead to a sub-optimal equilibrium, where the generator finds a pattern of fake examples with poor quality but absurdly high scores_false and the discriminator fails to produce natural scores_real that are higher than scores_false. This leads to an extremely low loss for the generator, but the discriminator gets stuck in a state with a very large loss. As such, we use an ordinary least square (OLS) version of the generator loss function, that forces the generator to generate fake samples with scores as close to real samples as possible. This prevents the aforementioned problem above, while ensuring that the generated examples are trained to be indistinguishable from the fake examples. Besides that, we extend the loss function described above to include other loss terms, forming the supervised component of the generator loss. These loss terms measure the OLS loss between the first three moments and the covariance matrices of the real and fake price paths, and the first three moments of scores_real and scores_false. Again, these loss terms evaluate the whole distribution of price paths for each unique condition embedding.

In summary, the modified semi-supervised loss function is defined as below, where $a_1, b_1, b_{1,\times}, c_1, a_2, b_2$, and c_2 are hyperparameters:

$$Loss_{gen} = output_{OLS} + scores_{OLS},$$

where

$$output_{OLS} = a_1 \times mean((mean(paths_{fake}) - mean(paths_{real}))^2) + b_1 \times$$
$$mean((stdev(paths_{fake}) - stdev(paths_{real}))^2) + b_{1,\times} \times mean((offdiag(covar(paths_{fake}) -$$
$$offdiag(covar(paths_{real}))^2) + c1 \times mean((skew(paths_{fake}) - skew(paths_{real}))^2)$$

$$scores_{OLS} = a_2 \times (mean(scores_{fake}) - mean(scores_{real}))^2 + b_2 \times (stdev(scores_{fake}) -$$
$$stdev(scores_{real}))^2 + c_2 \times (skew(scores_{fake}) - skew(scores_{real}))^2$$

For each unique condition embedding, the loss function above will be applied on 10 real examples and all 30 fake examples.

4 Evaluation

Now that we have defined the model architecture and successfully trained the model. We can then evaluate the model's resulting outputs and compare them with benchmarks. We will go over the few Monte-Carlo benchmarks and evaluation metrics that we use, before we present the results.

4.1 Monte-Carlo Benchmarks

The three benchmarks we use are popular models used commonly in traditional Monte-Carlo simulations in finance, which are described as follows. Note that we simulate all these benchmark models with the antithetic variate approach.

4.1.1 Geometric Brownian Motion (GBM)

GBM is the standard and most used dynamic to model underlying asset prices, where the drift μ and volatility σ are constant within each simulation and scales proportionately to the asset price $\{S_t\}$. The GBM's analytical formulation is as below:

$$dS_t = \mu S_t dt + \sigma S_t dW_t, \ or \ S_t = S_0 exp\left(\left(\mu - \frac{\sigma}{2}\right)t + \sigma W_t\right).$$

To obtain values for μ and σ at each time point, we assume a "risk-neutral" setting that allows us to avoid the inaccurate estimate of mean so as to stabilize and improve the simulation results.

Keeping our 2-month window in mind, μ is set to be the 3-month market yield on US treasury securities, which we extracted from Yahoo!Finance via the symbol "DGS3MO". σ is set to the implied volatility of S&P500 at-the-money options with 60 days to expiry, which was previously extracted from the Bloomberg LIVE.

4.1.2 Constant Elasticity of Variance (CEV)

The CEV model (see [2]) is a local volatility model that has similar dynamics to GBM except for the γ variable, given by

$$dS_t = \mu S_t dt + \sigma S_t^\gamma dW_t,$$

where γ is the elasticity of variance and is also constant for each simulation. γ is a useful feature because when $\gamma < 1$, this produces a phenomenon where the volatility increases as its price falls, which is common in equity markets as the stock's leverage ratio increases.

As CEV's closed-form analytical solution is complex and difficult to compute, we calibrated this model's parameters μ, σ, and γ using the S&P500 option market prices previously retrieved from OptionsDx. Calibration was made possible using Python's scipy.optimize.minimize() function, where we minimize the squared loss between model prices and market prices.

4.1.3 Heston's Model

Heston's model (see [5]) is a stochastic volatility model, whereby the variance v follows another stochastic process with long-run average variance θ at the rate of reversion κ, and volatility (vol of vol) ξ. The stock price and its variance dynamics are given as follows:

$$dS_t = \mu S_t dt + \sqrt{v_t} S_t dW_t^S,$$

$$dv_t = \kappa(\theta - v_t)dt + \xi \sqrt{v_t}\left(\rho dW_t^S + \sqrt{1 - \rho^2}dW_t^v\right),$$

where $\{W_t^S\}$ and $\{W_t^v\}$ are two independent standard Brownian motions. We utilize Python's QuantLib library that contains classes and methods to calibrate all Heston's parameters given market option prices.

4.2 Visualizations

We develop some helpful visualizations to show how different the distribution of the simulations from the CC-TTS-GAN is, compared to the benchmark models. Here, we only use the testing period from Apr 2022 to May 2024 and thus all synthetic simulations here by our CC-TTS-GAN are out-of-sample. For each time-point within this testing period, we run simulations for 2 months ahead of that time-point and compare them with the actual price path realized after 2 months.

4.2.1 Time-Series Simulations Charts

Figures 2, 3, 4, 5 are some time-series plots of the CC-TTS-GAN and benchmark's simulations relative to the real price paths. In each plot, the black curve is the actual realized price path, while the grey curve is the condition used, which is the implied volatility in this case. For readability, we only show one set of simulations (in colors) for each 20 days, and we cut each simulation down from 2 months to 1 month. Note that our benchmark model simulations are cut off after Jan 2024, as OptionsDx only provides options chain data up until Dec 2023. Nonetheless, we are still able to gain much valuable insights from these visual comparisons.

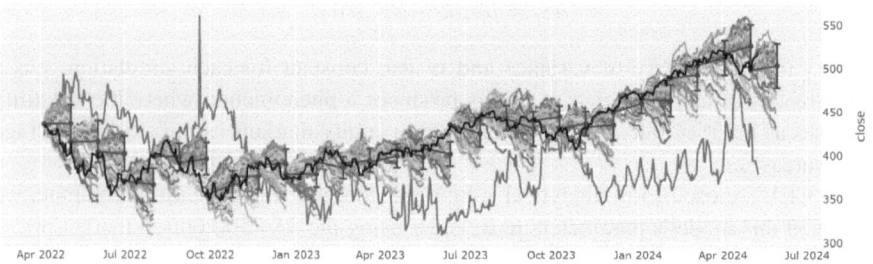

Fig. 2. Simulation Plot (CC-TTS-GAN vs Real)

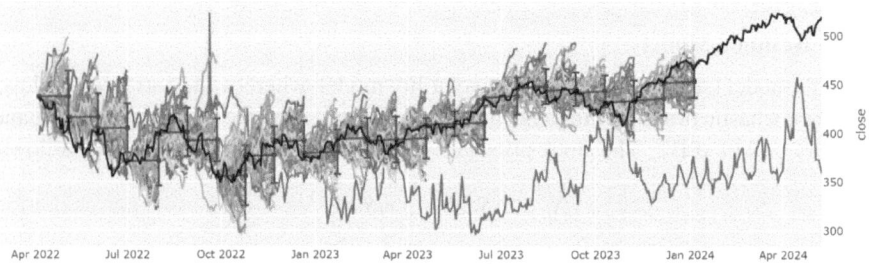

Fig. 3. Simulation Plot (GBM vs Real)

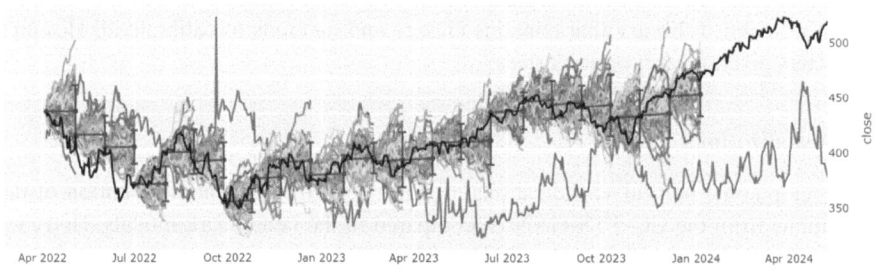

Fig. 4. Simulation Plot (CEV vs Real)

As can be observed, the CC-TTS-GAN simulations tend to exhibit less volatility relative to the benchmark models. This could be seen as an improvement as the benchmark models (usually calibrated based on implied volatility or market prices) tend to always over-estimate volatility of the time series. The CC-TTS-GAN does not seem to exhibit such weaknesses and can provide more precise distributions. Moreover, the CC-TTS-GAN is also able to provide relatively accurate directional convictions of predicted price movements, while the CEV and Heston's models barely deviate from risk-neutrality even when their drift parameter is calibrated on market prices.

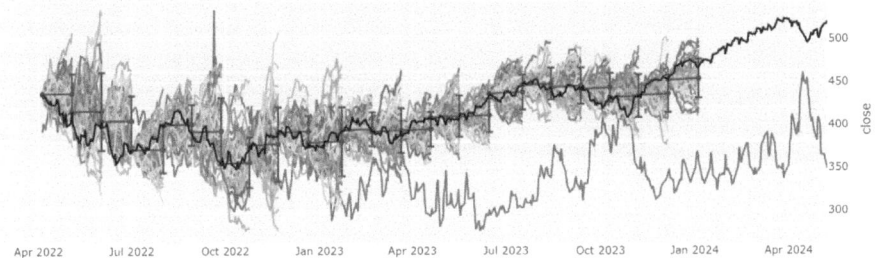

Fig. 5. Simulation Plot (Heston vs Real)

4.2.2 Scatter Plots after Dimensionality Reduction

To provide another perspective of the model results, it is common to leverage principal component analysis (PCA) and t-distributed Stochastic Neighbor Embedding (t-SNE) plots. To this end, we ought to reduce the dimensions of our 2-month-long simulations from 42 (2 months = approximately 42 trading days) to 2 using PCA. The resulting scatter plots are shown in Figs. 6 and 7, where red points refer to real price paths, blue points refer to synthetic simulations by the CC-TTS-GAN, and green points refer to the GBM benchmark simulations. Note that displaying all the five groups of data (real + synthetic + 3 benchmarks) would render the scatter-plot unreadable, which is why we omitted the CEV and Heston's data points in those two figures.

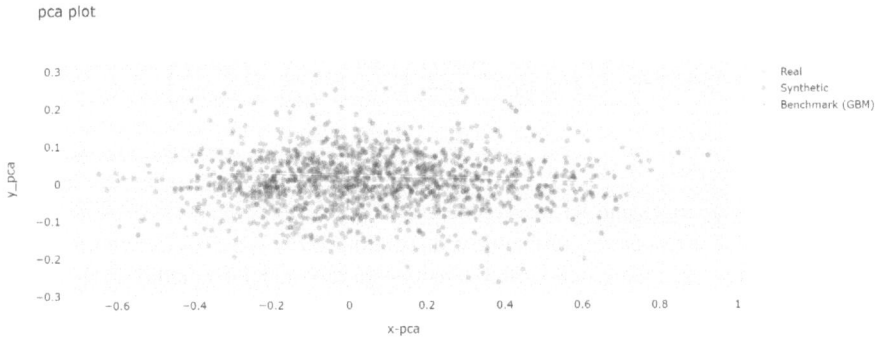

Fig. 6. PCA Scatter Plot

We can observe that the CC-TTS-GAN can better model the tails by noticing that red and blue points are more aligned with around the extreme values while less green points emerge in those regions. The coverage of the blue and red points are consistent. However, reducing 42 dimensions to 2 leads to much information loss, and thus we can hardly conclude which generated distribution is the closest to the real distribution. As such, we will introduce evaluation metrics, which will provide more objective measures of performance for all our simulations.

tsne plot

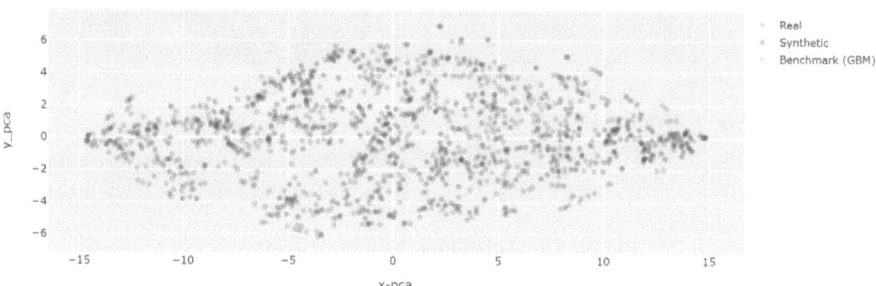

Fig. 7. t-SNE Scatter Plot

4.3 Distance Metrics

4.3.1 Jensen–Shannon (JS) divergence

JS-divergence (see [9]) is based on the Kullback–Leibler (KL) divergence. Both are metrics that measure the distance between two distributions. However, unlike KL-divergence, JS-divergence is symmetric, which means $D_{JS}(P\|Q) = D_{JS}(Q\|P)$ but in contrast, $D_{KL}(P\|Q) \neq D_{KL}(Q\|P)$, where P and Q are two respective probability distributions. Its symmetry makes JS-divergence a more robust metric. Their formulas are given as follows:

$$D_{KL}(P\|Q) = \int P(x)\log\frac{P(x)}{Q(x)}dx,$$

$$D_{JS}(P\|Q) = \frac{1}{2}D_{KL}\left(P\|\frac{P+Q}{2}\right) + \frac{1}{2}D_{KL}\left(Q\|\frac{P+Q}{2}\right).$$

As per the formulas, when $P(x)$ and $Q(x)$ are identical distributions, $D_{JS}(P\|Q) = D_{JS}(Q\|P) = D_{KL}(P\|Q) = D_{KL}(Q\|P) = 0$. The more different the distributions are, the larger the JS-divergences and KL-divergences are.

To estimate the respective probability distributions $P(x)$ and $Q(x)$, where x is a price path, we leverage Gaussian kernel density estimation (KDE) methods. However, when price path x has many dimensions (42, in our case, for a 2-month simulation), the estimation of $P(x)$ and $Q(x)$ becomes numerically unstable and may fail. As such, we use PCA once again to reduce x's dimensions from 42 to 10, before running the KDE, which simplifies the estimation process.

4.3.2 Fréchet Inception Distance (FID)

FID is another distance metric introduced by [6], traditionally used to measures the difference between simulated images produced by GANs by looking at their means and covariance matrices, as below:

$$FID = \|\mu_{real} - \mu_{fake}\|^2 + Trace\left(\Sigma_{real} + \Sigma_{fake} - 2(\Sigma_{real}\Sigma_{fake})^{\frac{1}{2}}\right).$$

We intend to apply this metric on our simulated and real price paths.

4.3.3   Root Mean Square Error (RMSE) of Mean

In our evaluation, this means the RMSE between the real price path, and the mean of generated price paths (averaged at each time point).

4.3.4   Dynamic Time Warping (DTW) Distance of Mean

Instead of measuring Euclidean distances like RMSE, DTW is a dynamic programming algorithm that measures differences in temporal sequences by aligning them in a way that minimises the distance between corresponding points, even if the sequences vary in speed or timing, as shown in Fig. 8. This makes DTW distance suitable for time-series data like price paths, where we do not expect exact alignment between two series, but a general pattern. In our evaluation, we use the DTW distance of mean, which is the DTW distance between the real price path, and the mean of generated price paths (averaged at each time point).

After training the model, we ran 20 trials of evaluation. Each trial contains 50 simulations for every available time-point in the testing period, which results were compared with benchmarks. These results were then aggregated across the 20 trials and summarised below (lower scores are better).

Table 2. Summary of Distance Metric Results

	GBM	CEV	Heston	CC-TTS-GAN
JS-Divergence (PCA)	**2.2182**	**2.3317**	**1.9198**	1.5211
JS-Divergence (t-SNE)	**0.2189**	**0.2323**	0.1409	0.1368
FID	**0.0047**	0.0028	0.0020	0.0035
RMSE of Mean	0.0596	0.0596	0.0596	0.0607
DTW of Mean	0.2027	0.2027	0.2027	0.1930

As per Table 2, CC-TTS-GAN outperforms all benchmarks in 3 out of the 5 distance metrics, and remains comparable with the benchmarks for the other 2 metrics. Values with underlines are the lowest (best) within that metric. For each distance metric and benchmark, we also conducted the hypothesis test below:

- H_0: CC-TTS-GAN's score = benchmark's score
- H_1: CC-TTS-GAN's score > benchmark's score

Values in bold are where the hypothesis test rejects the null hypothesis H_0.

Besides the direct comparison, we observe the following about the results in Table 2. First, as expected, Heston's model tends to achieve lower distance metrics than GBM and CEV, as it accounts for the stochastic component of volatility and thus reflects more accurate modelling of price dynamics. Second, note that the RMSE and DTW distance scores for the 3 benchmarks may appear identical, but they are actually different if we look at their values at order of magnitude -16. These extremely close scores are expected and can be explained from Figs. 3, 4, 5, where we mentioned that the benchmark simulations barely deviate from risk-neutrality, implying that their drift coefficients (and hence their means) are approximately at the risk-free rate. This is unlike the scores for CC-TTS-GAN simulations, which do not adhere to risk-neutrality and reflect more realistic market dynamics.

4.4 Runtime Comparisons

Lastly, we also compare the runtimes of each benchmark and CC-TTS-GAN simulations. Using Python's tqdm library, we were able to record the number of iterations per second, where each iteration refers to a completed set of 100 simulated price paths at one time-point. We selected 100 time-points and thus the results are averaged across 100 iterations.

This runtime includes any calibration process for the benchmark models. However, we exclude the model training time of the CC-TTS-GAN, as modern state-of-the-art GAN models always require long training times for high-quality generations. Below is a summary list of what the runtime includes for each model's simulation process:

(1) *GBM: Risk-neutral; no calibration. Simply substitute treasury yields and implied volatility data for each time-point, then simulate.*
(2) *CEV: Calibrated with market prices via manual simulations for each time-point, then simulate.*
(3) *Heston: Calibrated with market prices via Quantlib for each time-point, then simulate.*
(4) *CC-TTS-GAN: Input noise vector and condition embeddings into trained model, then forward propagation to retrieve simulations.*

Table 3 shows the results we obtained (more iterations are better).

Table 3. Summary of Runtime

	GBM	CEV	Heston	CC-TTS-GAN
#Iterations per Second	76.38	2.54	5.27	18.83

Each model's simulation process is different from one another, and thus we see a large variety of values. GBM is the fastest as it did not go through any calibration, while CEV is the slowest as we performed manual simulations for each calibration. CC-TTS-GAN achieves a decent runtime, ranking second, although GBM still far outperforms it. This shows that once trained, CC-TTS-GAN can provide time savings as it skips the calibration process, which is a benefit over certain models that have no closed-form solution and require complex calibration.

5   Conclusion

Overall, our evaluation results showcase how the CC-TTS-GAN framework could produce accurate simulations conditioned on some given continuous market variables, while being on par with benchmark models. The CC-TTS-GAN's capability of continuous conditional generation makes it a potentially value-adding tool to modern industry users.

The points below summarize the main definitive benefits of the CC-TTS-GAN over the benchmark models:

(1) **Model-free setting**: *No assumptions on risk-neutrality and price dynamics, which seems to tie benchmark models down.*

(2) **Skip calibration process**: *Once trained, allows for much faster simulations over complex models.*

(3) **Flexibility of input data**: *Complex models require very specific options chain data, which are often expensive or difficult to source. The CC-TTS-GAN can accept any input data as conditions, if they have some predictive power over price paths.*

(4) **Do not overestimate volatilities**: *The implied volatilities exhibited by benchmark models are empirically always inflated.*

(5) **Reflect future drift of price path more accurately**: *Benchmark models are heavily restrained by risk-neutrality, and thus their drift is mostly set near the risk-free rate.*

However, we also noted some weaknesses of the CC-TTS-GAN relative to the benchmark models:

(1) *Needs longer training time: Comes with all deep learning models, though this can be sped up with GPU resources eg. Kaggle's GPU T4 x2 shortened our training times to just between 30min and 1h.*

(2) *Needs lots of training data: Training any GAN would require vast amounts of training examples, which is particularly problematic in finance, where the further back our price data is, the more different the price dynamics are due to the inherent non-stationary.*

(3) *Needs hyperparameter tuning: The hyperparameters in our complex supervised loss function (as per Sect. 3.3.2) would need to be retuned each time if we wish to apply our model on a different financial asset.*

(4) *Needs feature engineering: This comes with the flexibility of input data – the user would need to find and engineer relevant features to be used as conditions. In addition, due to the way data points are sampled in training (as per Sect. 3.3.1), the user needs to ensure that the condition embeddings are evenly distributed, though this can be easily done with scikit-learn's preprocessing.QuantileTransformer() on Python.*

5.1 Future Work

There are still many aspects of exploration and improvement for the CC-TTS-GAN. One aspect is to implement and evaluate the model across other equities and ETFs, so that we can further validate its performance and robustness.

Besides, we believe the CC-TTS-GAN can be vastly improved if we make further modifications to its architecture. We could consider using discriminator/critic model that evaluates a single distribution of time series as a whole, which removes the need for the complex supervised loss function in Sect. 3.3.2 and avoids any tedious hyperparameter tuning. However, evaluating distributions as a whole would mean we need a lot more data, which brings us to our next point.

The modified model could be applied on higher frequency prices and signals, where there are more training examples in a shorter timeframe. Higher frequency prices may also exhibit greater stationarity due to short-run mean reversions, volatility clusters and lower likelihood of consistent trending, all of which are likely to allow us to sample a larger number of training examples which dynamics are more similar to one another, improving model performance.

References

1. Arjovsky, M., Chintala, S., Bottou, L.: Wasserstein generative adversarial networks. In: Proceedings of the 34th International Conference on Machine Learning (ICML 2017), vol. 70. PMLR (2017)
2. Cox, J.C.: The constant elasticity of variance option pricing model. J. Portfolio Manage. **23**, 15–17 (1996). https://doi.org/10. 3905/jpm.1996.015
3. Xin Ding, Yongwei Wang, Zuheng Xu, William J. Welch, and Z. Jane Wang. 2023. Continuous Conditional Generative Adversarial Networks: Novel Empirical Losses and Label Input Mechanisms. IEEE Transactions on Pattern Analysis and Machine Intelligence 45, 7 (July 2023), 8143–8158. https://doi.org/10.1109/tpami. 2022.3228915
4. Goodfellow, I., et al.: Generative adversarial nets. Adv. Neural Inf. Process. Syst. **27**. NIPS (2014)
5. Heston, S.L.: A closed-form solution for options with stochastic volatility with applications to bond and currency options. Rev. Financ. Stud. **6**(2), 327–343 (1993). https://doi.org/10. 1093/rfs/6.2.327
6. Heusel, M., Ramsauer, H., Unterthiner, T., Nessler, B., Hochreiter, S.: GANs trained by a two time-scale update rule converge to a local nash equilibrium. Adv. Neural Inf. Process. Syst. **30**. NIPS (2017)
7. Kolesnikov, A., et al.: An image is worth 16x16 words: transformers for image recognition at scale. In: Proceedings of the 9th International Conference on Learning Representations. ICLR (2021)
8. Li, X., Metsis, V., Wang, H., Ngu, A.H.H.: TTS-GAN: a transformer-based time-series generative adversarial network. In: Michalowski, M., Abidi, S.S.R., Abidi, S. (eds.) Artificial Intelligence in Medicine. AIME 2022. Lecture Notes in Computer Science(), vol. 13263, pp. 133–143. Springer, Cham (2022). https://doi.org/10.1007/978-3-031-09342-5_13
9. Menéndez, M.L., Pardo, J.A., Pardo, L., Pardo, M.C.: The jensen-shannon divergence. J. Franklin Inst. **334**(2), 307–318 (1997). https://doi.org/10.1016/s0016-0032(96)00063-4
10. Vaswani, A., et al.: Attention is all you need. Adv. Neural Inf. Process. Syst. **30**. NIPS (2017)
11. Vuletić, M., Prenzel, F., Cucuringu, M.: Fin-GAN: forecasting and classifying financial time series via generative adversarial networks. Quant. Financ. **24**(2), 175–199 (2024). https://doi. org/10.1080/14697688. 2023.2299466

Author Index

A
Akwiwu-Uzoma, Chukwuebuka 9
Apriyanto, Gaguk 32

B
Bonelli, Marco I. 137

C
Cambria, Erik 20
Chang, Chun-Hsien 111
Chen, Li Jun 176
Chiang, Ying Jen 164

D
Dahiya, Liza 20
Do, Ba-Lam 67
Durodola-Tunde, Kehinde 9

G
Ghosh, Rudra Chandra 88

H
Ho, Horstann Rui Yao 191
Hwang, Min-Shiang 111

J
Jemai, Jaber 1

K
Karim, Maisha 76
Kumar, Rajdeep 88
Kwok, Ron Chi-Wai 164

L
Lebea, Khutso 56
Li, Steven 176

Lin, Cheng-Ying 111
Liu, Beier 151
Liu, Jiahao 137
Liu, Yuxin 122
Lukianchenko, P. 99

M
Ma, Yu 20
Malik, Abdul 32
Manro, Rohan 20
Mao, Rui 20

N
Nghiem, Viet-Thang 67
Nwachukwu, Chukwuemeka 9

O
Ovuehor, Samuel 9

P
Pavlova, A. 99
Podkorytov, Roman 164
Pun, Chi Seng 191

R
Rahman, Mohammed Mizanur 76

S
Sangweni, Sphamandla 56
Setyowati, Rini 32
Sharma, Nitin 88
Siau, Keng 122
Sihwahjoeni, 32
Singh, Ganesh Bahadur 88
Sun, Mingjun 151

K.-W. Huang et al. (Eds.): ICFT 2024, CCIS 2437, pp. 207–208, 2025.
https://doi.org/10.1007/978-981-96-3811-6

T
Tran, Thi-Huong   67

W
Wang, Hoi-Hei   45
Wang, Runyu   122
Wang, Zhaojie   45

Y
Yang, Cheng-Ying   111

Z
Zhu, Haiyun   151
Zotov, G.   99
Zuhroh, Diana   32

The manufacturer's authorised representative in the EU is Springer
Nature Customer Service Centre GmbH, Europaplatz 3, 69115 Heidelberg,
Germany. If you have any concerns regarding our products, please
contact ProductSafety@springernature.com

Printed and bound by CPI Group (UK) Ltd, Croydon, CR0 4YY
27/04/2026
02097586-0002